Statistical Analysis of Climate Series

Helmut Pruscha

Statistical Analysis of Climate Series

Analyzing, Plotting, Modeling, and Predicting with R

 Springer

Helmut Pruscha
Ludwig-Maximilians-Universität
München
Germany

ISBN 978-3-642-43087-9   ISBN 978-3-642-32084-2   (eBook)
DOI 10.1007/978-3-642-32084-2
Springer Heidelberg New York Dordrecht London

Preface

Abstract The topic of this contribution is the statistical analysis of climatological time series. The data sets consist of monthly (and daily) temperature means and precipitation amounts gained at German weather stations. Emphasis lies on the methods of time series analysis, comprising plotting, modeling and predicting climate values in the near future. Further, correlation analysis (including principal components), spectral and wavelet analysis in the frequency domain and categorical data analysis are applied.

Introduction Within the context of the general climate discussion, the evaluation of climate time series gains growing importance. Here we mainly use the monthly data of temperature (mean) and precipitation (amount) from German weather stations, raised over many years. We analyze the series by applying statistical methods and describe the possible relevance of the results. First the climate series (annual and seasonal data) will be considered in their own right, by employing descriptive methods. Long-term trends—especially the opposed trends of temperature in the nineteenth and twentieth century—are statistically tested. The auto-correlations, that are the correlations of (yearly, seasonally, monthly, daily) data, following each other in time, are calculated before and after a trend or a seasonal component is removed. In the framework of correlation analysis, we use principal components to structure climate variables from different stations. We also formulate well-known folk (or country) sayings about weather in a statistical language and check their legitimacy.

The notion of auto-correlation leads us to the problem, how to model the evolution of the underlying data process. For annual data, we use ARMA-type time series models, applied to the differenced series, with a subsequent residual analysis to assess their adequacy. For the latter task, GARCH-type models can be employed. In the present text, predictions of the next outcomes are understood as forecasts: The prediction for time point $t + 1$ is strictly based on information up to time t only (thus parameters must be estimated for each t anew). The ARMA-type modeling is compared with (left-sided) moving averages by using a goodness-of-fit criterion calculated from the squared residuals.

Guided by the modeling of annual data, we similarly proceed with monthly data. Here, it is the detrended series, to which we fit an ARMA-model. With this method, the yearly seasonality can correctly be reproduced.

Daily records on temperature reveal a seasonal component—as known from monthly data, such that the adjusting of the series is advisable. We study a spatial effect, namely the cross-correlation between five German stations. Half of the daily precipitation data consists of zeros; here we are led to logistic regression approaches, to categorical data analysis and—with repect to heavy precipitation—to event-time analysis.

We continue with analyses in the frequency domain. Periodograms, spectral density estimations, and wavelet analyses are applied to find and trace periodical phenomena in the series.

Then, we present two approaches for predicting annual and monthly data, which are quite different from those based on ARMA-type models, namely growing polynomials and sin-/cos-approximations, respectively. Further, the one-step predictions of the preceding sections are extended to l-step (i. e. l-years) forecasts. This is done by the Box and Jenkins and by the Monte Carlo method. Finally, specific features of temperature and of precipitation data are investigated by means of multiple correlation coefficients.

The numerical analysis is performed by using the open-source package

R   [cran.r-project.org].

An introductory manual as for instance the book of Dalgaard (2002) is useful. The R codes are presented within complete programs. We have two kinds of comments. If comments should appear in the output, they are standing between "..." signs. If they are only directed to the reader of the program and should be ignored by the program, they begin with the ♯ sign. Together with the read.table(...) command in program R 1.1, the programs are ready to run. Optionally the sink(...) command in program R 2.1 can be employed (to divert the output to an external file). The index lists the R commands with the page of their first occurrence.

This book addresses

- Students and lecturers in statistics and mathematics, who like to get knowledge about statistical methods for time series (in a wide sense) on one side and about an interesting and relevant field of application on the other
- Meteorologists and other scientists, who look for statistical tools to analyze climate series and who need program codes to realize the work in R.

Programs, which are ready to run, and data sets on climatological series (both provided on the author's homepage) enable the reader to perform own exercises and allow own applications.

www.math.lmu.de/∼pruscha/   Helmut Pruscha

Contents

Chapter 1
Climate Series

Basic informations on four German weather stations and on the climate series, analyzed in the following chapters, are presented. The series consist of monthly temperature and monthly precipitation records. From these records, we derive seasonal and yearly data.

1.1 Weather Stations

Our data sets stem from the following four weather stations; further information can be found in Table 1.1 and in the Appendix A.1.

Bremen. The city Bremen lies in the north German lowlands, 60 km away from the North Sea. Weather records started in 1890.
Source: www.dwd.de/ (*Climate Environment, Climatological Data*).

Hohenpeißenberg. The mountain HoherPeißenberg (989 m) is situated between Weilheim and Schongau (Bav.) and lies in the lee-area of the Alps. It is the place of weather recording since 1781.
Source: www.dwd.de/ (*Climate Environment, Climatological Data*).
Further Grebe (1957), Attmannspacher (1981).

Karlsruhe. The town lies in west Germany in the upper Rhine lowlands. Weather recording started in 1799, but stopped at the end of 2008.
Source: www.klimadiagramme.de (*Klima in Karlsruhe*).

Potsdam. Since 1893 we have weather records from this east German town near Berlin.
Source: http://saekular.pik-potsdam.de (*Klimazeitreihen*).

⎡**R 1.1**⎤ The climate data HohenT, HohenP etc. are supposed to be stored in the folder C:/CLIM in the format of text-files. Their form is reproduced in the Appendix A.1 (additionally, separating lines - - - - - - - are used in A.1). The header consists of the variable names
Year dcly jan feb mar apr may jun jul aug sep oct nov dec Tyear
resp. Pyear. Dcly is the repetition of the December value of the last year (to have

H. Pruscha, *Statistical Analysis of Climate Series*,
DOI: 10.1007/978-3-642-32084-2_1, © Springer-Verlag Berlin Heidelberg 2013

Table 1.1 Survey of the four weather stations

Name	Height (m)	Geographical latitude	Geographical longitude	Start of temperature series	Start of precipitation series
Bremen	5	53°02′	08°47′	1890	1890
Hohenpeißenberg	977	47°48′	11°00′	1781	1879
Karlsruhe	112	49°02′	08°21′	1799	1876
Potsdam	81	52°23′	13°03′	1893	1893

the three winter months side by side). The data are loaded into the R program—according to the special application—by one or several of the following commands (T and Tp stand for temperature, P and Pr for precipitation).

```
hohenTp<- read.table("C:/CLIM/HohenT.txt",header=T)
hohenPr<- read.table("C:/CLIM/HohenP.txt",header=T)
karlsTp<- read.table("C:/CLIM/KarlsT.txt",header=T)
karlsPr<- read.table("C:/CLIM/KarlsP.txt",header=T)
potsdTp<- read.table("C:/CLIM/PotsdT.txt",header=T)
potsdPr<- read.table("C:/CLIM/PotsdP.txt",header=T)
```

and by analogy `bremenTp`, `bremenPr`.

1.2 Temperature Series

We have drawn two time series plots for each station: the annual temperature means (upper plot) and the winter means (lower plot). The meteorological winter covers the December (of the last year) and January, February (of the actual year). Winter data are often considered as an indicator of general climate change; but see Sect. 8.4 for a discussion. One finds the plots for

Bremen (1890–2010) in Fig. 1.1
Hohenpeißenberg (1781–2010) in Fig. 1.2
Karlsruhe (1799–2008) under the author's homepage
Potsdam (1893–2010) in Fig. 1.3.

The temperature strongly decreased in the last recorded year, i.e., in 2010, as it happens from time to time, for instance in the years 1987 and 1996 before.

R 1.2 Computation of some basic statistical measures that are sample size `length()`, mean value `mean()`, standard deviation `sqrt(var())`, correlation `cor()`.
To explain, how a user built `function()` operates, the computation is done first for the variable *yearly* temperature, then threefold—by means of the user function `printL`—for the three variables *yearly*, *winter*, *summer* temperature.

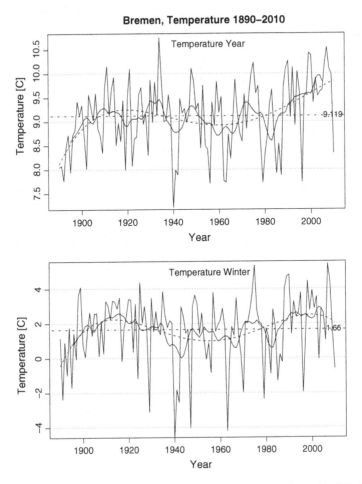

Fig. 1.1 Annual temperature means (*top*) and winter temperature means (*bottom*) in (°C), Bremen, 1890–2010; with a fitted polynomial of fourth order (*dashed line*), with centered (11-years) moving averages (*inner solid line*) and with the total average over all years (*horizontal dashed-dotted line*). The mean values for 2011 are 10.14 (year) and 0.33 (winter)

The two last commands `detach()` and `rm()` are omitted in the following R programs.

```
attach(hohenTp)

Y<- Tyear/100;            "annual temperature means in Celsius"
N<- length(Y); meanY<- mean(Y); sdY<- sqrt(var(Y))
rhoY<- cor(Y[1:(N-1)],Y[2:N])
c("N Years"=N, "Mean"=meanY, "StDev"=sdY, "Autocor(1)"=rhoY)

#------------------------------------------------------------
```

Table 1.2 Descriptive measures of the seasonal and the annual temperature data in (°C), for the four stations

	Bremen $n = 121$			Hohenpeißenberg $n = 230$		
	m	s	$r(1)$	m	s	$r(1)$
Winter	1.650	2.000	0.151	−1.366	1.730	0.076
Spring	8.449	0.988	0.175	5.571	1.322	0.165
Summer	16.769	0.973	0.111	14.259	1.079	0.208
Autumn	9.477	0.953	0.093	6.970	1.302	0.008
Year	9.119	0.749	0.350	6.359	0.845	0.296
	Karlsruhe $n = 210$			Potsdam $n = 118$		
	m	s	$r(1)$	m	s	$r(1)$
Winter	1.763	1.888	0.113	0.174	2.080	0.124
Spring	10.182	1.083	0.242	8.468	1.158	0.196
Summer	18.722	1.071	0.250	17.428	1.023	0.164
Autumn	10.192	1.029	0.046	8.921	1.074	0.081
Year	10.217	0.802	0.332	8.786	0.815	0.356

Mean m, standard deviation s, auto-correlation of first order $r(1)$

```
printL<- function(Y){
N<- length(Y); meanY<- mean(Y); sdY<- sqrt(var(Y))
rhoY<- cor(Y[1:(N-1)],Y[2:N])
#as last command of the function printL
c("N Years"=N, "Mean"=meanY, "StDev"=sdY, "Autocor(1)"=rhoY) }

Y<- Tyear/100;          "annual temperature means in Celsius"
printL(Y)

Y<- (dcly+jan+feb)/30;  "winter temperature means in Celsius"
printL(Y)

Y<- (jun+jul+aug)/30;   "summer temperature means in Celsius"
printL(Y)

detach(hohenTp)
rm(list=objects())              #remove all objects from workspace
```

In our data sets, the Dec. value of the last year is repeated in each new line (under the variable name dcly). If this is not the case, the winter temperature can be calculated by R commands as follows. Note that the first value for dcly is put artificially as the average of the first 10 dec values.

```
Y<- 1:N; Y[1]<- (mean(dec[1:10])+jan[1]+feb[1])/30
Y[2:N]<- (dec[1:(N-1)]+jan[2:N]+feb[2:N])/30
```

Table 1.2 offers the outcomes of some descriptive statistical measures that are mean value (m), standard deviation (s), auto-correlation of first order ($r(1)$). The

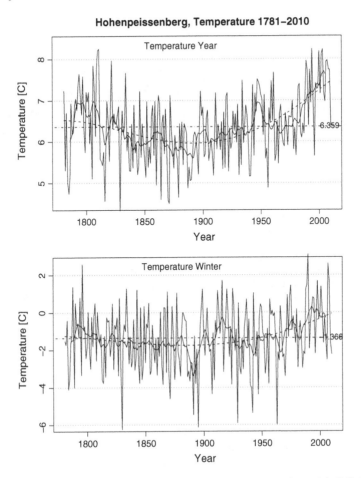

Fig. 1.2 Annual temperature means (*top*) and winter temperature means (*bottom*) in (°C), Hohen-peißenberg, 1781–2010; legend as in Fig. 1.1. The mean values for 2011 are 8.48 (year), which is the highest value since 1781, and −0.93 (winter)

latter describes the correlation of two outcomes (of the same variable) immediately following each other in time.

Discussion of the row *Year*: The annual mean values stand in a distinct order: Karlsruhe > Bremen > Potsdam > Hohenpeißenberg. However, their oscillations s are nearly of equal size ($\approx 0.8\,°C$ around the mean), and so are even the auto-correlations $r(1)$. That is, the correlation between the averages of two consecutive years amounts to 0.29...0.36. We will see below, how much therefrom is owed to the long-term trend of the series.

Discussion of the rows *Winter... Autumn*: The winter data have the largest oscillations s ($\approx 2\,°C$) and small auto-correlations $r(1)$. (Even smaller are the $r(1)$ values of the autumn data, signalizing practically zero correlation.) The time series plots of the winter series (especially the lower plots of Figs. 1.2, 1.3) reflect the s and $r(1)$

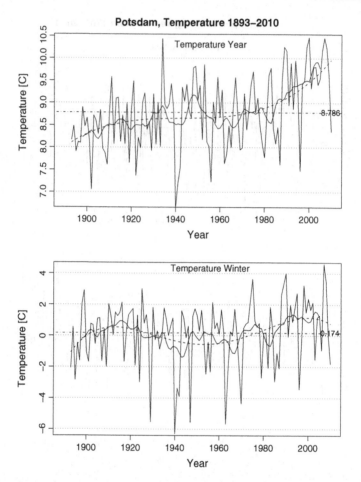

Fig. 1.3 Annual temperature means (*top*) and winter temperature means (*bottom*) in (°C), Potsdam, 1893–2010; legend as in Fig. 1.1. The mean values for 2011 are 10.14 (year) and −1.17 (winter)

values of the Table 1.2. In comparison with the upper plots of the annual means, they show a higher fluctuation and a less distinct trend, coming nearer to the plot of a pure random series (that is a series of uncorrelated variables).

R 1.3 Plot (by means of `plot()`) of annual temperature means, together with a fitted polynomial of order four and with centered (11-years) moving averages; see upper part of Fig. 1.2.

The polynomial is produced by the linear model commands `lm` and `predict` and enters the plot by `lines`.

The postscript file is stored under `C:/CLIM/HoTpYe.ps`.

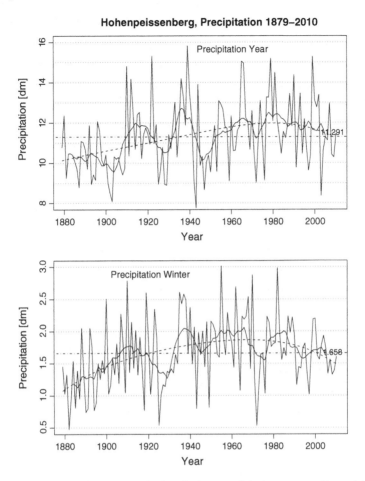

Fig. 1.4 Annual precipitation amounts (*top*) and winter precipitation amounts (*bottom*) in (dm), Hohenpeißenberg, 1879–2010; with a fitted polynomial of fourth order (*dashed line*), with centered (11-years) moving averages (*inner solid line*) and with the total average over all years (*horizontal dashed-dotted line*). The values for 2011 are 12.47 (year) and 1.36 (winter)

```
attach(hohenTp)
quot<- "Hohenpeissenberg, Temperature 1781-2010"; quot
postscript(file="C:/CLIM/HoTpYe.ps",height=6,width=16,horiz=F)

Y<- Tyear/100                              # annual means in Celsius
Ja<- Year-1800                             # to have smaller values
#fitting polynomial of order 4
J2<- Ja*Ja; J3<- J2*Ja; J4<- J3*Ja
tppol<- lm(Y~Ja+J2+J3+J4)

#centered (11-years) moving averages
```

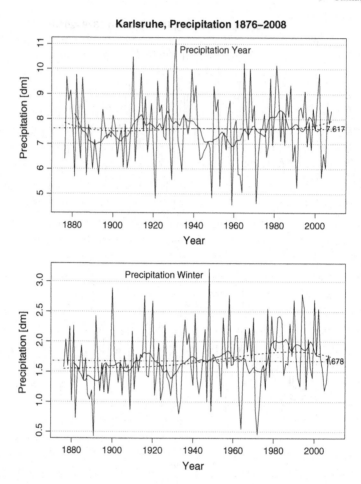

Fig. 1.5 Annual precipitation amounts (*top*) and winter precipitation amounts (*bottom*) in (dm), Karlsruhe, 1876–2008; same legend as in Fig. 1.4

```
N<- length(Y); p<- 10; m<- p/2
glD<- 1:N; su<- 1:N                          #glD, su vectors of dim N
for (t in (m+1):(N-m))
{su[t]<- 0
{ for (k in -(m-1):(m-1))  su[t]<-  su[t]+ Y[t+k] }}
for (t in (m+1):(N-m))
{glD[t]<- 0                                  #weight 1/2 at the margins
  glD[t]<- glD[t]+((Y[t-m]+Y[t+m])/2+su[t])/p}

ytext<- "Temperature [C]"; ttext<- "Temperature Year"
cabl<- c(4:8)                                #for horizontal lines
```

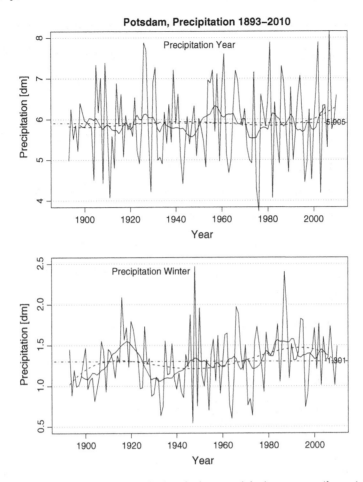

Fig. 1.6 Annual precipitation amounts (*top*) and winter precipitation amounts (*bottom*) in (dm), Potsdam, 1890–2010; same legend as in Fig. 1.4. The values for 2011 are 6.20 (year) and 1.30 (winter)

```
plot(Year,Y,type="l",lty=1,xlim=c(1780,2010),ylim=c(4.5,8.5),
  xlab="Year",ylab=ytext,cex=1.3)
title(main=quot); text(1880,8.4,ttext,cex=1.2)
abline(h=cabl,lty=3); abline(h=mean(Y),lty=4)
#print total mean with 3 digits into the plot:
text(2010,mean(Y),round(mean(Y),3),cex=0.8)

lines(Year,predict(tppol),lty=2)            #polynomial fitted
lines(Year[(m+1):(N-m)],glD[(m+1):(N-m)],lty=1)   #moving aver.

dev.off()                                   #output of the graphic
```

Table 1.3 Descriptive measures of the seasonal and the annual precipitation amount h in (mm) for the four stations

	Bremen $n = 121$			Hohenpeißenberg $n = 132$		
	h	s	$r(1)$	h	s	$r(1)$
Winter	152.4	51.7	−0.009	165.9	53.9	0.146
Spring	146.9	43.0	0.042	263.5	72.2	0.227
Summer	215.6	60.6	−0.113	454.2	94.3	−0.122
Autumn	168.1	50.1	−0.106	245.4	78.2	0.041
Year	682.8	106.7	0.052	1129.2	171.9	0.274
	Karlsruhe $n = 133$			Potsdam $n = 118$		
	h	s	$r(1)$	h	s	$r(1)$
Winter	167.9	54.5	−0.041	130.1	37.1	0.050
Spring	177.5	55.8	0.111	130.9	41.9	−0.021
Summer	227.7	69.5	−0.201	195.4	59.9	−0.017
Autumn	188.5	64.4	−0.013	133.8	43.7	−0.185
Year	761.7	135.3	0.009	590.5	96.0	−0.079

Standard deviation s, auto-correlation of first order $r(1)$

1.3 Precipitation Series

Again, we have drawn two time series plots for each station: the yearly precipitation amounts (upper plot) and the winter amounts (lower plot). One finds the plots for

Bremen (1890–2010) under the author's homepage
Hohenpeißenberg (1879–2010) in Fig. 1.4
Karlsruhe (1876–2008) in Fig. 1.5
Potsdam (1893–2010) in Fig. 1.6.

Table 1.3 offers the total precipitation amount (h) in (mm) height, standard deviation (s), and auto-correlation of first order ($r(1)$).

The *annual* precipitation amount at the mountain Hohenpeißenberg is nearly twice the amount in Potsdam. The oscillation values s stand in the same order as the amounts h. That is different from the temperature results in Table 1.2, where all four s values were nearly the same. Note that the precipitation scale has a genuine zero point, but the temperature scale has none (which is relevant for us).

The *winter* (Bremen: *spring*) is the season with the least precipitation (the smallest h) and with the smallest oscillation s (recall, that winter *temperature* had the largest s). Unlike the annual amounts, the winter amounts do not differ very much at the four stations.

While the precipitation series of winter and year in Bremen, Karlsruhe and Potsdam—with their small $r(1)$ coefficients—resemble pure random series, these series at Hohenpeißenberg however do not (see also Sects. 3.3, 4.4 and 8.4).

Chapter 2
Trend and Season

Polynomials, moving averages and straight lines—the latter two describe the decrease
and increase of temperature in the last two centuries—are considered. The warming
in the last 20 years is substantiated. The effect of auto-correlation on standard significance tests is discussed. The study of monthly data gives rise to introduce the notion
of a seasonal component and of seasonally adjusted data. Finally, we plot the course
of oscillation (fluctuation) of a climate variable and search for trends or patterns.

2.1 Trend Polynomials. Moving Averages

A trend component describes the long-term variation of a time series.
A comparatively rough and little sophisticated method is to fit polynomials (of lower
order) to the whole time series Y_t, $t = 1, \ldots, n$, of observed annual data. Here,
n is the number of years. See Figs. 1.1, 1.2, 1.3, 1.4, 1.5 and 1.6 for fourth-order
polynomials

$$p_t = b_0 + b_1 \cdot t + b_2 \cdot t^2 + b_3 \cdot t^3 + b_4 \cdot t^4, \quad t = 1, 2, \ldots, n. \qquad (2.1)$$

The residuals from the fitted polynomial are given by $e_t = Y_t - p_t$. A goodness-of-fit
measure is calculated from the mean sum of the *squared residuals* (MSQ) by

$$RootMSQ = \sqrt{\frac{1}{n} \sum_{t=1}^{n} e_t^2}.$$

The smaller the measure, the better the fit of the polynomial. Due to $\bar{e} \approx 0$ the
measure $RootMSQ$ is approximately equal to the standard deviation of the
residuals e_t. Table 2.1 shows that the $RootMSQ$- values decrease with increasing
order k. This decrease is very slow for precipitation and stronger for temperature.
For temperature at Hohenpeißenberg and in Karlsruhe, the biggest drop is from order

H. Pruscha, *Statistical Analysis of Climate Series*,
DOI: 10.1007/978-3-642-32084-2_2, © Springer-Verlag Berlin Heidelberg 2013

Table 2.1 Annual temperature means

	Bremen				Hohenpeißenberg			
	Temperature		Precipitation		Temperature		Precipitation	
k	R	r_1	R	r_1	R	r_1	R	r_1
1	0.719	0.30	1.05	0.02	0.824	0.26	1.64	0.21
2	0.717	0.29	1.05	0.02	0.757	0.12	1.62	0.19
4	0.681	0.21	1.03	−0.00	0.754	0.11	1.62	0.18
6	0.675	0.20	1.01	−0.04	0.737	0.06	1.59	0.16
	Karlsruhe				Potsdam			
	Temperature		Precipitation		Temperature		Precipitation	
k	R	r_1	R	r_1	R	r_1	R	r_1
1	0.790	0.31	1.35	0.01	0.733	0.21	0.95	−0.08
2	0.707	0.14	1.35	0.01	0.724	0.18	0.95	−0.08
4	0.692	0.11	1.34	0.00	0.712	0.15	0.95	−0.09
6	0.672	0.06	1.32	−0.05	0.702	0.14	0.95	−0.09

Order k of the polynomial and resulting goodness-of-fit R = RootMSQ. Further, the auto-correlation $r_1 = r_e(1)$ of the residual series e_t is listed

$k = 1$ (straight line) to order $k = 2$ (parabola)—more than from $k = 2$ to 4, 6. For temperature, the auto-correlations $r_e(1)$ of the residuals are distinctly positive, meaning that the fit p_{t+1} stays—by tendency—on the same side of the observed value as the fit p_t does. The same is true with precipitation only at Hohenpeißenberg.

The $r_e(1)$-values for precipitation in Bremen, Karlsruhe, and Potsdam are ≈ 0, but that was already the case with the $r(1)$-values of the original series Y, see Table 1.3.

Next, we compare the fitted polynomials (of order $k = 1, 2, 3$) for three stations. For the sake of comparability, we take 116 years (1893–2008) only and center the curves around $\bar{Y}$; that is, we are plotting in Fig. 2.1 the values $p_t - \bar{Y}, t = 1, \ldots, 116$. The curves run nearly identical over the 116 years. That is, the annual temperature means—when approximated by polynomials—run remarkably parallel at the three stations.

Further, Figs. 1.1, 1.2, 1.3, 1.4, 1.5 and 1.6 contain—as trend curves—*centered* moving averages m_t over $k = 11$ years. Putting $k = 2 * l + 1$, for estimating the trend m_t we form the time interval $[t - l, t + l]$ of k points, with the time point t as the center, and extend the average over the k years, but with weight $1/2$ for the endpoints; that is

$$m_t = \frac{1}{2 * l} \cdot \left[\frac{1}{2} \cdot Y_{t-l} + Y_{t-l+1} + \cdots + Y_t + \cdots + Y_{t+l-1} + \frac{1}{2} \cdot Y_{t+l} \right]. \quad (2.2)$$

Remark. The variables m_t or p_t, according to Eqs. (2.1) or (2.2), are predictions (interpolations) for Y_t. Note that they use information from observations before and after time point t. Let us call this approach the *standard regression approach* for

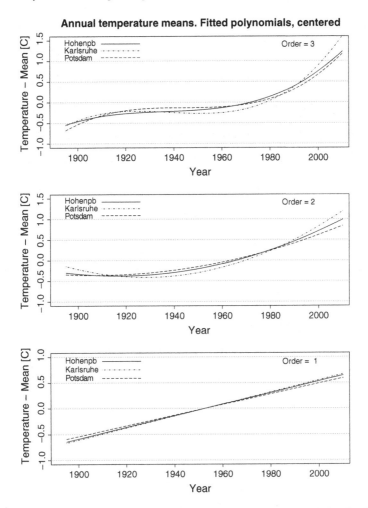

Fig. 2.1 Fitting polynomials of order 1, 2, and 3 over the years 1893–2008, each time for the three stations Hohenpeißenberg, Karlsruhe, Potsdam. Each curve is centered around the total mean $\bar{Y}$ for the station

prediction. Within the context of climatological time series (which are continuously updated) a *forecast* approach for prediction seems to be more appropriate. Here, for predicting Y_t, only observations *before* time point t are employed. This is—for instance—the case with left-sided moving averages, growing polynomials, or autoregressive algorithms, which will follow in Chaps. 4, 5, and 8.

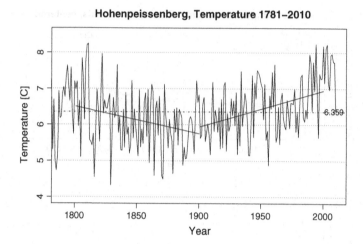

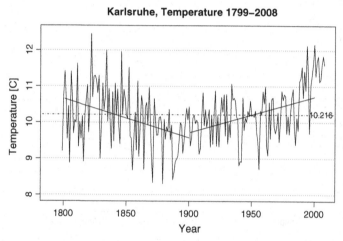

Fig. 2.2 Annual temperature means (°C) Hohenpeißenberg, 1781–2010 (*top*), Karlsruhe, 1799–2008 (*bottom*); with *straight lines* fitted for each century and with the total mean (*horizontal line*). Compare also Schönwiese (1995, Abb. 12)

2.2 Temperature: Last Two Centuries—Last Twenty Years

In this section, we study the long-term trend of temperature over the last two centuries. For this investigation, only the series of Hohenpeißenberg and of Karlsruhe are long enough. While temperature decreases in the nineteenth century, it increases in the twentieth century, see Fig. 2.2.

The regression coefficients (slopes) $b = b(Temp|Year)$ of the two—for each century separately fitted—straight lines $p_t = a + b \cdot t$ are tested against the hypothesis

Table 2.2 Statistical measures for the temperature (°C) of the last two centuries

	Hohenpeißenberg				
Period	n	Mean value	Standard deviation	Regression b*100	Correlation r	Test T
Nineteenth century	100	6.129	0.843	−0.763	−0.262	0.271
Twentieth century	100	6.445	0.747	1.006	0.390	0.423

	Karlsruhe				
Period	n	Mean value	Standard deviation	Regression b*100	Correlation r	Test T
Nineteenth century	100	10.114	0.845	−1.079	−0.370	0.398
Twentieth century	100	10.219	0.689	0.988	0.416	0.457

The regression coefficient b is multiplied by 100, r is the dimension-free version of b, T the test statistic (2.3)

of a zero slope. The level 0.01-bound for the test statistic T,

$$T = \frac{|r|}{\sqrt{1 - r^2}}, \quad r = b \cdot \frac{s(Year)}{s(Temp)}, \tag{2.3}$$

is $t_{98,0.995}/\sqrt{98} = 0.265$. Herein, the correlation coefficient r is the dimension-free version of b.

1. As Table 2.2 informs us, the negative trend in the 19th century and the positive trend in the twentieth century are statistically confirmed (at Hohenpeißenberg and in Karlsruhe). The test assumes uncorrelated residuals $e_t = Y_t - p_t$. This can be substantiated using the auto-correlation function of the e_t (not shown, but see Chap. 4 for similar analyses).

2. The total temperature means m_1 and m_2 of the two centuries do not differ very much from each other and from the total mean m of the whole series, see Table 2.2.

The increase of temperature in the 20th century is statistically significant in Potsdam, too. In Bremen, however, we have a nearly horizontal trend line over this time period (consult Fig. 2.3 and Table 2.3).

R 2.1 Plot of annual temperature means, together with straight lines fitted for the nineteenth and twentieth century separately, see Fig. 2.2 (bottom). The straight line is produced within the user function tempger (for lm and predict see also R 1.3). The output is written and stored on the file C:/CLIM/Tempout.txt.

```
attach(karlsTp)

postscript(file="C:/CLIM/KarlsT12.ps",height=6,width=20,horiz=F)
sink("C:/CLIM/Tempout.txt")          #Output on file Tempout.txt

quot<- "Karlsruhe,  Temperature 1799-2008"; quot
Y<- Tyear/100;    "annual means in Celsius"
cylim<-  c(8.0,12.5);   cabl<- c(8:12)
plot(Year,Y,type="l",lty=1,xlim=c(1790,2008),ylim=cylim,
   xlab="Year",ylab="Temperature [C]",cex=1.3)
```

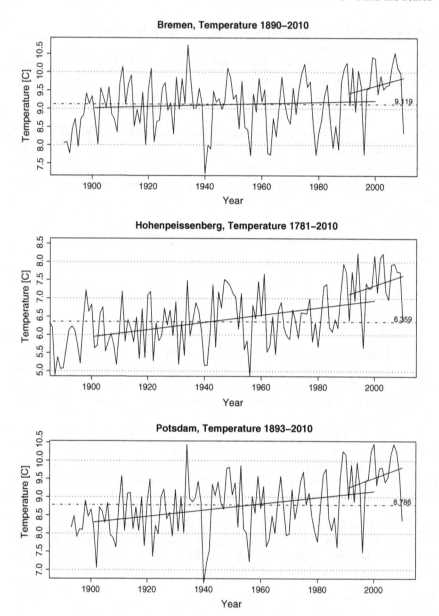

Fig. 2.3 Annual temperature means (°C). Bremen 1890–2010, Hohenpeißenberg 1781–2010 (here 1885–2010 is shown), Potsdam 1893–2010; with *straight line* fitted for the twentieth century. The *fitted line* for the 20 years 1991–2010 is also shown

Table 2.3 Statistical measures for the temperature (°C) of the last century and of the last 20 years 1991–2010

| Period | Bremen | | | | | | |
	n	Mean value	Standard deviation	Regression b*100	Correlation r	Test T	Upper limit
Twentieth century	100	9.117	0.738	0.198	0.077	0.077	
1991–2010	20	9.634	0.699	2.177	0.184		9.372 (16)

| Period | Hohenpeißenberg | | | | | | |
	n	Mean value	Standard deviation	Regression b*100	Correlation r	Test T	Upper limit
Twentieth century	100	6.445	0.747	1.006	0.390	0.423	
1991–2010	20	7.370	0.699	2.677	0.226		6.553 (17)

| Period | Potsdam | | | | | | |
	n	Mean value	Standard deviation	Regression b*100	Correlation r	Test T	Upper limit
Twentieth century	100	8.732	0.804	0.860	0.310	0.326	
1991–2010	20	9.542	0.730	2.919	0.237		9.066 (16)

The regression coefficient b is multiplied by 100, r is the dimension-free version of b, T the test statistic (2.3). The upper limit refers to the 99 % confidence interval (2.4); in brackets the number of years (out of 20) with a temperature mean above the upper limit

```
title(main=quot)
abline(h=cabl,lty=3);   abline(h=mean(Y),lty=4)
text(2008,mean(Y),trunc(mean(Y)*1000)/1000,cex=0.8)    #total mean

#-------------------------------------------------------------
tempger<-  function(Year,Y,A,B){  #compute and plot straight line
Y0<- Y[A:B]; Year0<- Year[A:B]
tpger0<- lm(Y0~Year0);     tpg0<- summary(tpger0)
lines(Year0,predict(tpger0),lty=1)              #plot fitted line
return(tpg0)                                     #return summary
}

"19th century"
Jbeg<- 2;    A1<- Jbeg+1; B1<- Jbeg+100;
tpg1<- tempger(Year,Y,A1,B1); tpg1                #print summary

"20th century"
A2<- Jbeg+101; B2<- Jbeg+200;
tpg2<- tempger(Year,Y,A2,B2); tpg2                #print summary

dev.off()
```

Output from R 2.1 Excerpt from results written on the file
C:/CLIM/Tempout.txt, for Karlsruhe, Temperature.

The square root of R-squared = 0.137 equals the absolute value 0.370 of the coefficient of correlation in Table 2.2. The t-value -3.945 divided by $\sqrt{98}$ equals the absolute value 0.398 of the test statistic T.

```
"19th century"
Call:  lm(formula = Y0 ~ Year0)
Coefficients:
              Estimate Std. Error t value  Pr(> t|)
(Intercept) 30.07826     5.0611     5.943  4.29e-08 ***
Year0       -0.010788    0.002735  -3.945  0.00015  ***
---
Signif. codes:  0 *** 0.001 ** 0.01 * 0.05 . 0.1    1
Residual standard error: 0.7894 on 98 degrees of freedom
Multiple R-squared: 0.137,  Adjusted R-squared: 0.1282
F-statistic: 15.56 on 1 and 98 DF,  p-value: 0.0001500
```

Effective Sample Size

When applying tests and confidence intervals to time series data, the effect of auto-correlation should be taken into account. To compensate, the sample size n is to be reduced to an *effective* sample size n_{eff}. As an example we treat the confidence interval for the true mean value μ of a climate variable, let us say the long-term temperature mean. On the basis of an observed mean value $\bar{y}$, a standard deviation s and an auto-correlation function $r(h)$, see Sect. 3.3 below, the $(1 - \alpha) * 100\%$ confidence interval (assuming a large n) is

$$\bar{y} - u_0 \cdot \frac{s}{\sqrt{n_{eff}}} \leq \mu \leq \bar{y} + u_0 \cdot \frac{s}{\sqrt{n_{eff}}}, \quad u_0 = u_{1-\alpha/2}, \tag{2.4}$$

with u_γ being the γ-quantile of the $N(0, 1)$-distribution, and with

$$n_{eff} = \frac{n}{1 + 2 \cdot \sum_{k=1}^{n-1}(1 - (k/n)) \cdot r(k)} \approx \frac{n}{1 + 2 \cdot \sum_{k=1}^{n-1} r(k)} \quad [n \text{ large}];$$

see Brockwell and Davis (2006, Sect. 7.1), von Storch and Zwiers (1999, Sect. 6.6). For an AR(1)-process with an auto-correlation $r = r(1)$ of first order we have to put $r(k) = r^k$, cf. Appendix B.3, and obtain

$$n_{eff} = n \cdot \frac{1 - r}{1 + r} \quad [n \text{ large}]. \tag{2.5}$$

The Last Twenty Years

We have the further result

3. The average m_3 over the last 20 years is significantly larger than the twentieth century mean m_2 (and larger than the total mean m too; 0.01 level). That is immediately confirmed by a two sample test, even after a correction, due to auto-correlation. The warming in the last two decades is well established by our data.

To make this result **3.** more explicit, we construct a 99% confidence interval around the long-term temperature mean μ acc. to (2.4) (where the auto-correlation is taken into regard). Then we count, how many of the last 20 yearly means lie above the upper limit.

Example Hohenpeißenberg: With $n = 230, r = 0.295, m = \bar{y} = 6.359, s = 0.844$ we are led by Eq. (2.5) to $n_{eff} = 125.21$ and thus to a 99% confidence interval $[6.165, 6.553]$.

For all three stations in Table 2.3, at least 16 of the last 20 yearly temperature means lie above the upper 99% confidence limit, reinforcing the result **3.** above. Among the exceptions are always the colder years 1991, 1996, 2010.

The *winter* temperatures show the same pattern, but in a weakened form. The fall and the rise of the straight lines are no longer significant (see result **1.**), at least 13 of the last 20 winter temperature means lie above the upper 99% limit of (2.4) (see result **3.**). So, the warming in the winter months of the last decades is not so strongly pronounced in our data.

2.3 Precipitation

The precipitation records start in the last quarter of the nineteenth century. To sketch their course over the last 120 years, we divide this time period into three intervals, namely

1891–1950 (Potsdam 1893–1950), 1951–1990 and 1991–2010 (Karlsruhe 1991–2008).

Then we calculate—for each time interval separately—the average of annual and of winter amounts. Further, a parabola is fitted over the whole 120 years. Figure 2.4 and Table 2.4 reveal a general increase of precipitation toward the second half of the last century. They show a drastic increase of the annual and the winter amounts from the first to the second time interval at Hohenpeißenberg (weaker in Bremen and in the winter data Karlsruhe), followed by a decrease to the third. The Table 2.4 reports the corresponding 2-sample t-test statistics. Note that the standard deviations are roughly between 1.0 (Po) and 1.7 (Ho) for the annual data, and between 0.4 and 0.5 for the winter data, cf. Table 1.3. Taking the maximal value of 0.27 for the auto-correlation into regard (and the correction formula in 2.2), the upper 5% bound for the absolute value of the t-test statistic is at most $t_{34-2,0.975} = 2.04$ (and at least $u_{0.975} = 1.96$, of course). Thus, statistically significant changes are:

- from the first to the second time interval at Hohenpeißenberg (annual and winter data) and in Bremen (annual data),
- from the second to the third interval at Hohenpeißenberg (winter data).

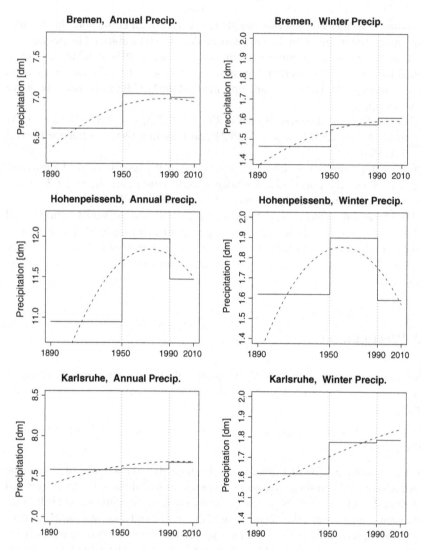

Fig. 2.4 Annual (*left*) and winter (*right*) precipitation amounts (dm), averaged over each of the 3 sections of the time period 1891–2010 (Bremen, Hohenpeißenberg), 1891–2008 (Karlsruhe). A parabola is fitted to the 120 yearly data. Notice that the y-axes on the *right* have the same range 1.4–2.0 (dm); the y-axes on the *left* have different ranges, but the ranges have the same width of 1.5 (dm)

Table 2.4 The table gives the annual and the winter precipitation amounts in (dm), for the three time intervals (1) 1891–1950, (2) 1951–1990, (3) 1991–2010, with the three mean values and with the two 2-sample t-test statistics for the changes from (1) to (2) and from (2) to (3)

Station	Mean 1	Mean 2	Mean 3	Test 1 → 2	Test 2 → 3
Bremen annual	6.623	7.057	7.014	2.23	−0.13
Bremen Winter	1.467	1.575	1.609	1.07	0.22
Hohenpb annual	10.95	11.97	11.48	2.92	−1.14
Hohenpb Winter	1.620	1.899	1.592	2.52	−2.42
Karlsruhe annual	7.582	7.599	7.683	0.06	0.22
Karlsruhe Winter	1.619	1.776	1.788	1.38	0.08
Potsdam annual	5.837	5.965	5.986	0.66	0.07
Potsdam Winter	1.249	1.336	1.384	1.11	0.47

2.4 Historical Temperature Variations

Statistical results are formal statements; they alone do not allow substantial statements on the earth warming. Especially, a prolongation of the upward lines of Figs. 2.2 and 2.3 would be dubious. An inspection of temperature variability of the last millenniums reveals that a trend (on a shorter time scale) could turn out as part of the normal variation of the climate system. See Schönwiese (1995), von Storch and Zwiers (1999).

Figure 2.5 shows temperature variability of the last 8,000 years, adopted from Schönwiese (1995), estimated by the method of oxygen-isotopes from Greenland's ice drill cores. Especially, we recognize distinctly cold and warm time periods, denoted by A–E in Fig. 2.5.

2.5 Monthly Values

The march of temperature and of precipitation over the 12 months of the year is plotted as histogram in Fig. 2.6. Hereby—for each specific month—the total average of n monthly values is calculated (n the number of years). In the case of temperature the histograms of the four stations (three are shown) show a rather similar form, with a somewhat lowered and compressed form for Hohenpeißenberg. In the case of precipitation, the wet months June and July at Hohenpeißenberg and the dry months February, March, and October in Potsdam attract attention.

According to Malberg (2007) the histogram of precipitation in Fig. 2.6 at the stations Hohenpeißenberg and Potsdam reflects more a continental (and less an oceanic) type of climate.

R 2.2 Six histograms of the total monthly averages for temperature and precipitation at three stations, see Fig. 2.6. Within the user function monthTP the (user)

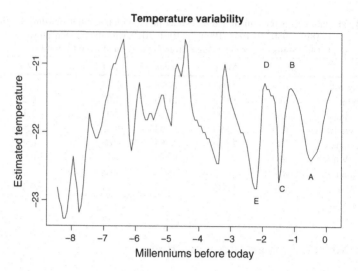

Fig. 2.5 Temperature variability of the last 8,000 years, qualitative curve; adopted from (Schönwiese (1995), Abb. 26). *A* (1500–1700), *B* (800–1000), *C* (450–800), *D* (200 BC–200 AD), *E* (1200 BC–600 BC)

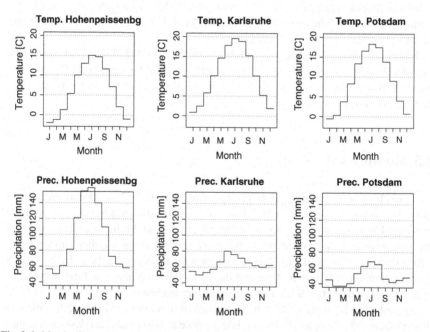

Fig. 2.6 March of temperature in °C (*top*) and precipitation in mm (*bottom*) over the calendar year, plotted for the three stations Hohenpeißenberg (until 2010), Karlsruhe (until 2008), Potsdam (until 2010). The truncated precipitation values for Hohenpeißenberg are: June 154.9, July 159.2

function plotTP is called. The latter produces a step function plot. Note that we put plm[13] = plm[12] with respect to the last (twelfth) step. The x-axis (side=1) with the initial letters is labeled by axis and labels. All six read.table commands of R 1.1 are needed.

```
postscript(file="C:/CLIM/MonthTP.ps",height=8,width=20,horiz=F)
par(mfrow=c(2,3),pty="s")                #2x3 pictures of square size

plotTP<- function(mo,ttext,cylim,tylab,cabl){
plmo<-c(mo,mo[12])              #plmo[13]: right corner of last step
x<- seq(0.5,12.5,by=1)
plot(x,plmo,type="s",                    #step function plot
   xlim=c(0.5,12.5),ylim=cylim,xaxt="n",xlab="Month",ylab=tylab)
axis(side=1,at=c(1:12),
   labels=c("J","F","M","A","M","J","J","A","S","O","N","D"))
title(main=ttext,cex=1.1);  abline(h=cabl,lty=3)
}
monthTP<- function(mon12,ttext,cylim,tylab,cabl){
mon12.mat<- as.matrix(mon12)                   #mon12 as matrix
mon12.me<- colMeans(mon12)                     #monthly means
plotTP(mon12.me,ttext,cylim,tylab,cabl)
}
#------------------------------------------------------------
cylim<- c(-2,20); tylab<- "Temperature [C]"
cabl<- c(0,5,10,15,20)
mon12<- data.frame(hohenTp[,3:14])/10          #select jan-dec
monthTP(mon12,"Temp. Hohenpeissenbg",cylim,tylab,cabl)
mon12<- data.frame(karlsTp[,3:14])/10
monthTP(mon12,"Temp. Karlsruhe",cylim,tylab,cabl)
mon12<- data.frame(potsdTp[,3:14])/10
monthTP(mon12,"Temp. Potsdam",cylim,tylab,cabl)

#----Similarly with precipitation----------------------------

dev.off()
```

To judge a temperature value in a specific month, we have to compare it with the value, which is predicted by the trend and by the seasonal component.

This comparison is illustrated by Fig. 2.7, which presents the 36 monthly temperature means Y_t of three succeeding years. The trend-component $\hat{m}_t$ is gained by building moving averages over 13 months, that is, by employing six preceding and six following months. The seasonal component $\hat{s}_t$ consists of the total averages of each month, as shown in the histogram of Fig. 2.6 (left, top)—centered at a mean value zero. The trend and season-component is then given by

$$\hat{Y}_t = \hat{m}_t + \hat{s}_t,$$

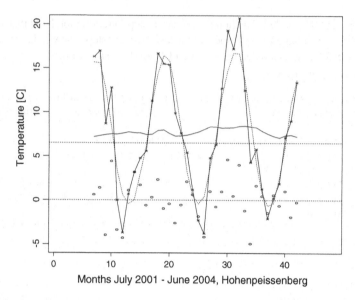

Fig. 2.7 The 36 monthly temperature means Y_t (×) at Hohenpeißenberg, July 2001–June 2004. In addition, with a trend-component (*inner solid line—*) and trend+season-component ($\cdots$), as well as residuals therefrom (o)

also called *prediction* for Y_t. The *residuals*

$$e_t = Y_t - \hat{Y}_t$$

reveal, for which months the trend- and seasonally adjusted temperature values are too high (then with a positive residual) or too low (then with a negative residual).

The "record summer" 2003 (months no. 30–32 in Fig. 2.7) is salient because of the above-average temperature values in June and August. Accordingly, the residual values are distinctly positive. Cold months (in relation to trend+season) were September, November, and December 2001, as well as especially October 2003—the latter with an extremely negative residual.

More sophisticated prediction/residual procedures for monthly data are presented in Chap. 5.

2.6 Oscillation in Climate Series

Besides the trend, it is also the oscillation (fluctuation) of a climatological series, in which we are interested. First, we want to visualize the oscillation of the *annual* temperature and precipitation values. To this end, we build moving 10-years blocks $[t-9, t]$, $t = 10, \ldots, n$, calculate for each block the standard deviation sd(t) = $\hat{\sigma}(t)$

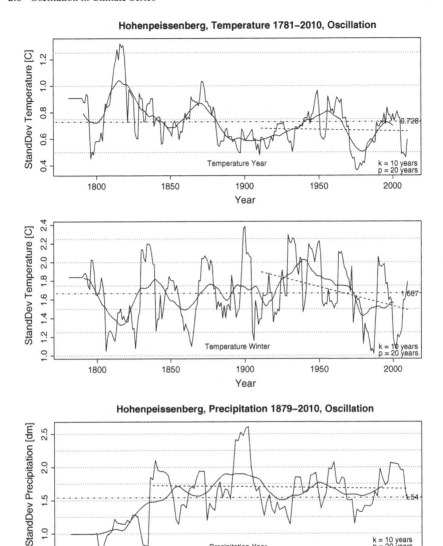

Fig. 2.8 Oscillation of annual climate values at Hohenpeißenberg. Standard deviation sd(t), calculated for 10-years blocks $[t - 9, t]$, plotted over years t. Further: smoothing by 20 years moving averages (*inner solid line*), *straight line* fit for the last 100 years (*dashed line*), and the total mean (*horizontal dashed-dotted line*). Shown are yearly and winter temperature means, yearly precipitation amounts (from *top* to *bottom*)

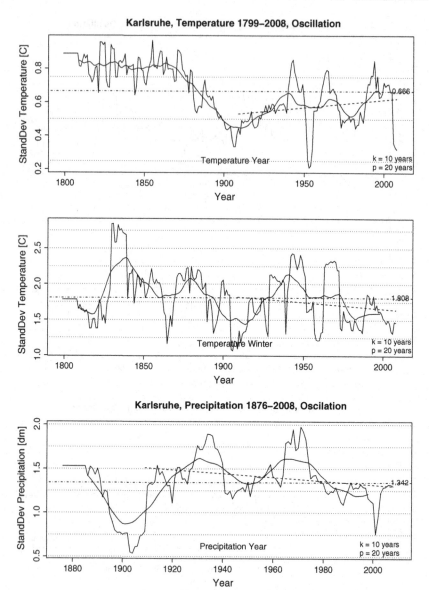

Fig. 2.9 Oscillation of annual climate values in Karlsruhe. Legend as for Fig. 2.8

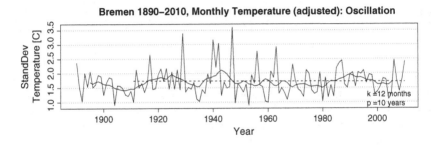

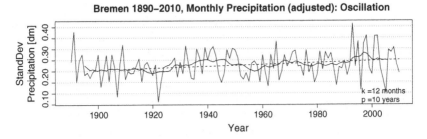

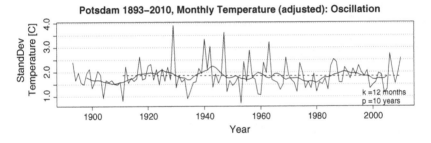

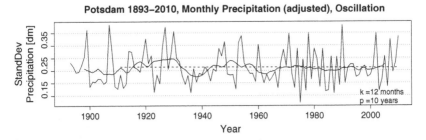

Fig. 2.10 Oscillation of (seasonally adjusted) monthly climate values, in Bremen and Potsdam. Standard deviation sd(t), calculated for calendar year *t*, plotted over the years *t*. Further: smoothing by 10 years moving averages (*inner solid line*), *straight line* fit for the last 100 years (*dashed line*)

and plot sd(t) over the years t. For Hohenpeißenberg (Fig. 2.8), Karlsruhe (Fig. 2.9), Bremen, and Potsdam (no Figs.), no definite common pattern can be detected, neither in the yearly nor in the winter data. Time periods with higher fluctuation follow those with lower fluctuation, without an apparent regularity and with little agreement between the stations. At least one could recognize a general lower oscillation around 1900 (except Fig. 2.8, middle). Further, perhaps against the expectation, the oscillation in the last 10 or 20 years is not very high. In Sect. 4.4, the oscillation in the annual precipitation series is analyzed by more sophisticated methods.

To quantify the oscillation of *monthly* climate values, we calculate for each calendar year the standard deviation sd(t), that is the standard deviation of the 12—seasonally adjusted—temperature means and precipitation sums, respectively. Once again, we plot sd(t) over the years t; see Fig. 2.10 for the stations Bremen and Potsdam. We do not discover clear-cut patterns, but with respect to temperature, we notice a good conformity of the Bremen and the Potsdam oscillation series sd(t).

The oscillation sd(t) shows no uniform trend over the last 100 years (see the straight line fit in Figs. 2.8, 2.9 and 2.10). The sign of the slopes differ between the four stations, and that is true for yearly and for winter temperature and precipitation, as well as for monthly precipitation. Only in the cases of monthly temperature we have an uniformly decreasing tendency (but the negative coefficients of slope are not significantly different from zero).

Chapter 3
Correlation: From Yearly to Daily Data

Scatterplots and correlation coefficients are defined for a bivariate sample $(x_1, y_1), \ldots, (x_n, y_n)$, where two variables, x and y, are measured n-times, each time at the same object or at comparable objects. When considering a whole set of variables, a matrix of pairwise correlations is established. Based on such a correlation matrix, the multivariate procedure of principal components can introduce a structure into the set of variables.

As special case the auto-correlation coefficient is considered, where x and y are the same variable, but taken at different time points. The effect of seasonal and trend components on auto-correlation is studied. We deal with the question, what the auto-correlation tells us when making predictions for the next observation. In this context, we also try to formulate folk- or country-sayings about weather in a statistical language and to check their legitimacy.

3.1 Auto-Correlation Coefficient

How strong is an observation at time point t (named x) correlated with the observation at the succeeding time point $t + 1$ (named y)? That is, we are dealing with the case, that x and y are the same variable (e.g., temperature Tp) but observed at different time points, symbolically

$$x = Tp(t), \quad y = Tp(t + 1).$$

The scatterplot of Fig. 3.1 (left) presents the 12*230 monthly temperature means at Hohenpeißenberg. The corresponding correlation coefficient is $r = r(1) = 0.79$; thus, the auto-correlation of monthly temperature (at Hohenpeißenberg) amounts to 0.79. The large value is owed to the seasonal effects, i.e., to the course of the monthly temperatures over the calendar year. It contains, so to say, much redundant information.

H. Pruscha, *Statistical Analysis of Climate Series*,
DOI: 10.1007/978-3-642-32084-2_3, © Springer-Verlag Berlin Heidelberg 2013

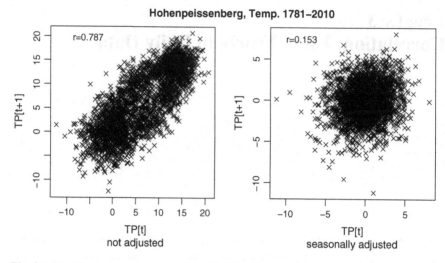

Fig. 3.1 Monthly temperature means TP = Y (°C). Scatterplots $Y(t+1)$ over $Y(t)$ with $n =$ $12*230 - 1$ points; *left* Original (not adjusted) variables, with correlation $r = 0.79$; *right* Variables after seasonal adjustment (i.e., after removal of monthly total averages), with correlation $r = 0.15$

Table 3.1 Auto-correlation $r(1) = r(Y_t, Y_{t+1})$ for climate variables (Hohenpeißenberg), without (in parenthesis) and with adjustment

	Temperature		Precipitation	
Succession	n	$r(Y_t, Y_{t+1})$	n	$r(Y_t, Y_{t+1})$
Year → succeeding Year	229	(0.296) 0.116	131	(0.274) −0.186
Winter → succeed. Wi	229	(0.076) 0.013	131	(0.146) −0.006
Summer → succeed. Su	229	(0.208) 0.104	131	(−0.122) −0.172
Winter → succeed. Su	229	(0.168) 0.100	131	(0.222) 0.171
Summer → succeed. Wi	229	(0.057) −0.014	131	(−0.023) −0.098
Month → succeed. Mo	2,759	(0.787) 0.153	1583	(0.379) 0.013
Day → succeeding Day	1,460	(0.932) 0.825	1460	(0.271) 0.250

In order to adjust, we first calculate 12 seasonal effects by the total averages for each month,

$$m_{jan}, \ldots, m_{dec},$$ together forming the seasonal component.

Figure 2.6 shows the seasonal component for three stations in the form of histograms. Then, we build *seasonally adjusted* data by subtracting from each monthly temperature mean the corresponding seasonal effect. The scatterplot of Fig. 3.1 (right) is based on these 12*230 adjusted monthly means, leading to the correlation coefficient $r = 0.15$. This is much smaller than the $r = 0.79$ from above for the non-adjusted case.

Tables 3.1, 3.2, and 3.3 offer auto-correlations $r(1) = r(Y_t, Y_{t+1})$ of climate variables Y for two successive time points. We deal with the variables

Y = yearly, quarterly, monthly, daily temperature, and precipitation.

Table 3.2 Auto-correlation $r(1) = r(Y_t, Y_{t+1})$ for climate variables (Karlsruhe), without (in parenthesis) and with adjustment

Succession	Temperature			Precipitation		
	n	$r(Y_t, Y_{t+1})$		n	$r(Y_t, Y_{t+1})$	
Year → succeeding Year	209	(0.332)	0.110	132	(0.009)	0.005
Winter → succeed. Wi	209	(0.113)	0.060	132	(−0.041)	−0.082
Summer → succeed. Su	209	(0.250)	0.064	132	(−0.201)	−0.230
Winter → succeed. Su	209	(0.175)	0.121	132	(0.104)	0.127
Summer → succeed. Wi	209	(0.119)	0.052	132	(−0.084)	−0.067
Month → succeed. Mo	2,519	(0.811)	0.197	1,595	(0.071)	0.029
Day → succeeding Day	1,460	(0.962)	0.867	1,460	(0.162)	0.157

Table 3.3 Auto-correlation $r(1) = r(Y_t, Y_{t+1})$ for climate variables (Potsdam), without (in parenthesis) and with adjustment

Succession	Temperature			Precipitation		
	n	$r(Y_t, Y_{t+1})$		n	$r(Y_t, Y_{t+1})$	
Year → succeeding Year	117	(0.356)	0.149	117	(−0.079)	−0.087
Winter → succeed. Wi	117	(0.124)	0.054	117	(0.050)	−0.025
Summer → succeed. Su	117	(0.164)	−0.083	117	(−0.017)	−0.039
Winter → succeed. Su	117	(0.068)	0.008	117	(0.187)	0.230
Summer → succeed. Wi	117	(0.125)	0.105	117	(0.041)	0.079
Month → succeed. Mo	1,415	(0.818)	0.276	1,415	(0.091)	0.005
Day → succeeding Day	1,460	(0.956)	0.857	1,460	(0.149)	0.142

The $r(1)$ coefficients for *day* were gained from 365*4 consecutive daily temperature and precipitation records of the years 2004–2007, see Sect. 6.1 and Appendix A.3.

R 3.1 Correlations of quarterly temperatures, after removal of a polynomial trend of order 4. This is done simultaneously for the 4 seasons Wi, Sp, Su, Au by using cbind. Note that variables A1<- A[1:(n-1)] and A2<- A[2:n] have a time-lag of 1 year; cor(varlist) answers with pairwise correlations between the members of varlist, in form of a (symmetrical) matrix.

```
attach(hohenTp)

n<- length(Year); options(digits=3)
Wi<- dcly+jan+feb; Sp<-mar+apr+may
Su<- jun+jul+aug;  Au<- sep+oct+nov      #no averaging necessary
Quar<- cbind(Wi,Sp,Su,Au)               #binding Wi,Sp,Su,Au together
                                        #Quar is a n x 4 matrix
"----Residuals from polynomials(4)-trend----"
Ja<- Year-1800;  Ja2<- Ja*Ja; Ja3<- Ja2*Ja; Ja4<- Ja2*Ja2
Quares<- Quar-predict(lm(Quar~Ja+Ja2+Ja3+Ja4))       #residuals
```

```
"1 | 2   refers to preceding | succeeding year"
Quares1<- cbind(Quares[(1:(n-1)),(1:4)])
Quares2<- cbind(Quares[(2:n),(1:4)])

Quares<- cbind(Quares1,Quares2)   #Quares is a (n-1) x 8 matrix
colnames(Quares)<- c("Wires1","Spres1","Sures1","Aures1",
                     "Wires2","Spres2","Sures2","Aures2")
cor(Quares)             #cross tabulation of pairwise correlations
```

| Output from R 3.1 | Cross tabulation of correlation coefficients.

Examples: cor(Wires1,Sures1) refers to winter and to the direct following summer; cor(Wires1, Sures2) to winter and to the summer of the next year.

```
"----Residuals from polynomials(4)-trend----"
        Wires1 Spres1 Sures1 Aures1 Wires2 Spres2 Sures2 Aures2
Wires1   1.000  0.116  0.100 -0.077  0.013  0.080  0.028  0.039
Spres1   0.116  1.000  0.162  0.166  0.076  0.069  0.126  0.100
Sures1   0.100  0.162  1.000  0.214 -0.014  0.128  0.104 -0.046
Aures1  -0.077  0.166  0.214  1.000  0.077  0.021  0.184 -0.099
Wires2   0.013  0.076 -0.014  0.077  1.000  0.121  0.101 -0.074
Spres2   0.080  0.069  0.128  0.021  0.121  1.000  0.160  0.170
Sures2   0.028  0.126  0.104  0.184  0.101  0.160  1.000  0.215
Aures2   0.039  0.100 -0.046 -0.099 -0.074  0.170  0.215  1.000
```

Besides the auto-correlation $r(1)$ of the non-adjusted variables (put in parenthesis) we present the $r(1)$ coefficient of the adjusted variables without parenthesis. Herein, adjustment refers to the removal

- of a trend component, more precisely, of a polynomial of order 4 (for each variable separately) in the case of *year*, *quarter*, and *day*. In the latter case, the polynomial was drawn over the 365 days of the calendar year, see Fig. 6.2
- of the seasonal component in the case of *month*.

Note that the non-adjusted temperature variables do not have negative auto-correlations (showing persistence), but some precipitation variables have (showing a switch-over effect).

In the following, we discuss exclusively the outcomes for the adjusted series that are the figures of Tables 3.1, 3.2, 3.3 not in parenthesis.

Temperature: As to be expected, the auto-correlation of the *daily* data is large. Smaller are those in the case of *month*, *year*, *quarter*. The monthly auto-correlations are larger than the yearly and the yearly are (with one exception) larger than the quarterly values.

Precipitation: Only at the mountain Hohenpeißenberg the auto-correlation of *yearly* data differs distinctly from zero. Here, the precipitation series has more inner structure than the series of Karlsruhe or Potsdam; see also the complements (Sect. 8.4). Completely different from the temperature situation, the auto-correlations of the *daily* precipitation data are—perhaps against expectations—comparatively small and that of the *monthly* data are nearly negligible.

What is the relevance of a particular $r(1)$ value, when we are at time t and the immediately succeeding observation (at time $t + 1$) is to be predicted? This will be discussed in Sect. 3.4.

3.2 Multivariate Analysis of Correlation Matrices

In the next step, a spatial aspect is included in our analysis. We consider the climatological variables *temperature* (Tp) and *precipitation* (Pr) as well as the five stations

Aachen (A), Bremen (B), Hohenpeissenberg (H), Karlsruhe (K), Potsdam (P)

(3.1)

in the years 1930–2008, see Appendix A.2. First, a 10×10 matrix of pairwise correlations is established. Let the 10 variables be denoted by TpA, PrA, ..., TpP, PrP. The 10×10 correlation matrix consists of four parts: the correlations between the five temperature variables (upper left) and between the five precipitation variables (lower right); further the cross-correlations between them (upper right and—symmetrically— lower left). In the latter two parts, we have mostly negative values and much smaller absolute values than in the first two parts.

	TpA	TpB	TpH	TpK	TpP	PrA	PrB	PrH	PrK	PrP
TpA	1.000	0.877	0.924	0.916	0.912	-0.018	-0.025	-0.160	-0.098	-0.020
TpB	0.877	1.000	0.774	0.812	0.939	-0.053	-0.058	-0.132	-0.092	-0.039
TpH	0.924	0.774	1.000	0.885	0.884	0.042	0.010	-0.253	-0.159	-0.030
TpK	0.916	0.812	0.885	1.000	0.874	0.104	0.043	-0.017	-0.054	0.044
TpP	0.912	0.939	0.884	0.874	1.000	0.004	-0.026	-0.112	-0.147	-0.069
PrA	-0.018	-0.053	0.042	0.104	0.004	1.000	0.608	0.466	0.509	0.579
PrB	-0.025	-0.058	0.010	0.043	-0.025	0.608	1.000	0.414	0.462	0.684
PrH	-0.160	-0.132	-0.253	-0.017	-0.112	0.466	0.414	1.000	0.430	0.432
PrK	-0.098	-0.092	-0.159	-0.054	-0.147	0.509	0.462	0.430	1.000	0.483
PrP	-0.020	-0.039	-0.030	0.044	-0.069	0.579	0.684	0.432	0.483	1.000

In order to summarize the information on correlation matrices and to structure the set of variables, we employ *principal component* analysis. For this multivariate procedure one may consult Morrison (1976), Hartung and Elpelt (1995), Fahrmeir et al. (1996). A short outline of this analysis goes as follows.

We start with our p observation variables, now denoted by $x_1, \ldots, x_p$; to each belongs an observation vector of length n (denoted by $x_1, \ldots, x_p$, too). In our case, we have $p = 10$ (later also $p = 5$) and $n = 79$. We assume that the vectors x_j are already *standardized* (mean 0, variance 1). As usual, we arrange these p vectors of length n in the form of an $n \times p$ data matrix X, i.e., $X = (x_1, x_2, \ldots, x_p)$. From this data, we derive the $p \times p$ correlation matrix $R = (X^\top \cdot X)/(n - 1)$. Let $\lambda_1 \geq \cdots \geq \lambda_p$ be the p positive eigenvalues of the matrix R and $a_1, \ldots, a_p$ the corresponding (orthogonal) eigenvectors,

$$R \cdot a_j = \lambda_j a_j, \quad j = 1, \ldots, p,$$

Fig. 3.2 Principal component analysis; temperature and precipitation in the years 1930–2008, at five stations A,B,H,K,P as in (3.1). The loadings of the observation variables are plotted in the plane, spanned by the first two components. Analysis is performed with the ten variables TpA, TpB, TpH, TpK, TpP (all lying in the *lower right corner*) PrA, PrB, PrH, PrK, PrP (*upper left corner*)

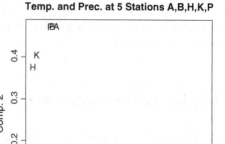

the vectors a_j normalized to 1. Now we build certain linear combinations of the p observation variables: the p vectors y_j of length n, defined by

$$y_j = X \cdot a_j, \quad j = 1, \ldots, p,$$

are called *principal components* (sometimes: principal factors). They are uncorrelated, with $\mathrm{Var}(y_j) = \lambda_j$ for each $j = 1, \ldots, p$. The value y_{ji} is the *jth factor score* for case i, $i = 1, \ldots, n$. The p eigenvectors a_j are arranged in the form of a $p \times p$ matrix Λ,

$$\Lambda = (a_1, a_2, \ldots, a_p),$$

the elements $\Lambda_{kj} = a_{jk}$ are called *loadings*. The loading a_{jk} (multiplied by $\sqrt{\lambda_j}$) equals the correlation between the (standardized) observation variable x_k and the principal component y_j. This fact serves as basis for the interpretation of the loadings.

First, we apply principal component analysis to the 10×10 correlation matrix, which is shown above. For each of the 10 climate variables TpA, PrA, ..., TpP, PrP, the first two components a_{1k}, a_{2k} of the loadings are plotted in Fig. 3.2. As it was to be expected, the five temperature variables and the five precipitation variables are lying strictly apart. Therefore, we apply the analysis to each of the two sets of variables separately, i.e., first to the upper left and then to the lower right 5×5 submatrix of the 10×10 correlation matrix.

Table 3.4 brings the 5×5 loading matrix Λ in the case of the five *temperature* variables. The first component can be comprehended as a general factor of magnitude. The second to fifth component describes differences between the stations: the second distinguishes Bremen (-0.71) on one side and Hohenpeißenberg (0.51), Karlsruhe (0.35) on the other. This is also expressed by Fig. 3.3 (left). The third component differentiates between Hohenpeißenberg (-0.58) and Karlsruhe (0.78). We will come back to this interpretation immediately.

Table 3.4 Principle component analysis of the five temperature variables (Tp) at the stations A,B,H,K,P as in (3.1)

Variable	Component 1	Component 2	Component 3	Component 4	Component 5
TpA	0.458	0.143	−0.019	−0.776	0.409
TpB	0.435	−0.709	0.056	−0.155	−0.530
TpH	0.442	0.513	−0.581	0.145	−0.427
TpK	0.444	0.349	0.779	0.258	−0.094
TpP	0.456	−0.304	−0.229	0.536	0.600
Standard deviations	2.126	0.516	0.344	0.261	0.165
Proportions of variance	0.904	0.053	0.024	0.014	0.005

The loading matrix Λ is given, together with the standard deviations (that are the square roots of the eigenvalues) and the proportions of variance, for each of the five components

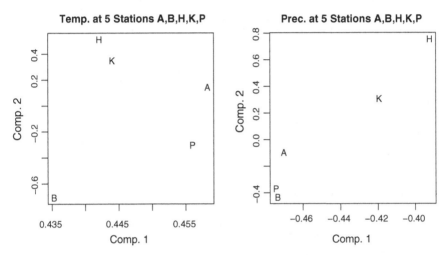

Fig. 3.3 Principal component analysis; temperature and precipitation in the years 1930–2008, at five stations A, B, H, K, P as in (3.1). The loadings of the observation variables are plotted in the plane, spanned by the first two components. *Left* Analysis with the five temperature variables TpA, TpB, TpH, TpK, TpP. *Right* Analysis with the five precipitation variables PrA, PrB, PrH, PrK, PrP (the first component with a negative sign)

Table 3.5 presents the analogous results for the five *precipitation* variables. Here, the second component differentiates between (Bremen, Potsdam) and (Hohenpeißenberg, Karlsruhe), see also Fig. 3.3 (right), the third between Hohenpeißenberg and Karlsruhe.

In Fig. 3.4, the first two factor scores y_{1i}, y_{2i} of the 79 cases (years) are plotted, for the five temperature variables (left) and for the five precipitation variables (right). For selected cases, numerical values of factor scores 1, 2, and 3 are given below.

Table 3.5 Principle component analysis of the five precipitation variables (Pr) at the stations A,B,H,K,P as in (3.1)

Variable	Component 1	Component 2	Component 3	Component 4	Component 5
PrA	−0.470	−0.096	0.000	−0.859	0.180
PrB	−0.473	−0.434	0.172	0.156	−0.730
PrH	−0.392	0.756	0.506	0.125	−0.049
PrK	−0.419	0.308	−0.834	0.172	−0.071
PrP	−0.474	−0.369	0.137	0.440	0.653
Standard deviations	1.744	0.817	0.747	0.650	0.558
Proportions of variance	0.608	0.134	0.112	0.085	0.062

See legend to Table 3.4. The first component appears with a negative sign

```
            Temperature              Precipitation
 No   Score1 Score2 Score3     Score1 Score2 Score3
 [1]   0.966 -0.338  0.162     -1.314  0.121 -1.380
 [2]  -2.570 -1.048  0.315     -1.108  0.105 -2.332
 [3]  -0.251 -1.047 -0.164      0.762  0.390  0.190
 [4]  -2.158 -0.897  0.340      1.889  1.493  0.571
 [5]   2.863 -1.015 -0.247      2.634  0.090  0.228
 [6]  -0.029 -0.965  0.750     -0.900 -0.056 -0.174
 [7]  -0.365  0.104  0.448     -0.692  1.281  0.474
 [8]   0.149  0.365  0.003     -0.337  0.537  0.650
 [9]   0.270 -0.630 -0.347      0.319  0.307 -0.175
[10]  -0.753 -0.455  0.101     -2.621  1.417  0.168
 ....                  ....
[46]   1.187 -1.100  0.120      1.893  0.976 -0.876
 ....                  ....
[70]   3.059 -0.443  0.445      0.163  2.801  0.180
[71]   4.011  0.343  0.278     -0.505  0.735  0.373
[72]   1.385  0.480  0.403     -2.040 -0.253  0.005
[73]   3.063  0.763  0.021     -4.055 -1.088  0.061
[74]   2.944  1.215  0.046      3.275 -0.778 -0.200
[75]   1.374  0.158  0.236      0.057 -0.969  0.367
[76]   1.635  0.046  0.394      0.865 -0.528  0.854
[77]   3.293  0.220 -0.018      0.693  0.575 -0.940
[78]   3.814 -0.074  0.134     -2.600 -0.925  0.803
[79]   2.747  0.018  0.090     -0.168 -0.430 -0.812
```

We will discuss some cases.

Temperature: Case No. 71 (year 2000) lies at the right border of Fig. 3.4 (left plot) with a maximal score 1 value of 4.01, but the score 2 is near zero (0.34). Accordingly, the temperature means of the year 2000 lie above the average—for all five stations (see the data set in Sect. A.2).

Case No. 46 (1975) is situated at the bottom of that figure, with a score 2 of −1.10. Accordingly in the year 1975, the temperature in Bremen is above the average, while the contrary is the case at Hohenpeißenberg.

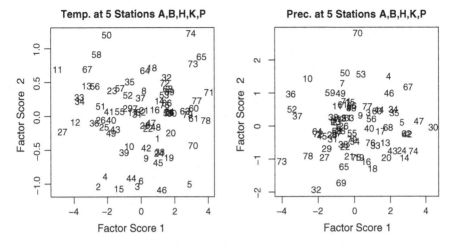

Fig. 3.4 Principal component analysis; temperature and precipitation in the years 1930–2008 at five stations A, B, H, K, P as in (3.1). The 79 cases (years) are plotted in the plane, spanned by the first two factor scores. *Left* Analysis with the five temperature variables TpA, TpB, TpH, TpK, TpP. *Right* Analysis with the five precipitation variables PrA, PrB, PrH, PrK, PrP (the first component with a negative sign)

Case No. 74 (2003), at the upper border, has Hohenpeißenberg (and Karlsruhe) far above the average, but Bremen only slightly.

Case No. 6 (1935) possesses a relatively large value of score 3. The temperature mean at Hohenpeißenberg lies below the average, that of Karlsruhe exceeds it.

Precipitation: Case No. 73 (2002) lies at the left border of Fig. 3.4 (right plot) with an extreme negative score 1 value of -4.05. Correspondingly, all five precipitation amounts of the year 2002 lie above the average (see Sect. A.2).

Case No. 70 (1999), at the upper border of that figure, has the largest score 2 value (2.80). The precipitation amount in Bremen in this year 1999 is below, the amount at Hohenpeißenberg far above the average.

Case No. 2 (1931), with minimal negative score 3 value of -2.33, differentiates the amounts at Hohenpeißenberg and Karlsruhe: The former lies below, the latter far above the average.

R 3.2 Principal component analysis with five temperature variables
TpA, TpB, TpH, TpK, TpP.

The data set Years5 can be found in Appendix A.2. After building the correlation matrix (cor), loadings (loadings) and factor scores (scores) are extracted from principal components (princomp) and are plotted; compare Figs. 3.3 (left) and 3.4 (left).

```
All5TP<- read.table("C:/CLIM/Years5.txt",header=T)
attach(All5TP)

postscript(file="C:/CLIM/All5T.ps",height=12,width=16,horiz=F)
par(mfrow=c(2,1),pty="s")                         #two square plots

quot<- "Temp. at 5 Stations A,B,H,K,P"; quot
all5T<-cbind(TpA,TpB,TpH,TpK,TpP)
txt<- c("A","B","H","K","P")
"Correlation matrix"; cor(all5T)
"Principal components"
all5T.pca<- princomp(all5T,cor=T);        summary(all5T.pca)
load5T<- -loadings(all5T.pca)            #minus sign for convenience
print(load5T,cutoff=0.01)        #print only loadings |.| > 0.01
x<- load5T[,1]; y<- load5T[,2]          #Comp 1, Comp 2 (out of 5)
plot(x,y,type="n",xlab="Comp. 1",ylab="Comp. 2")
text(x,y,txt,cex=0.9); title(main=quot,cex=0.8)
x<- -all5T.pca$scores[,1]; y<- -all5T.pca$scores[,2]     #dim 79
plot(x,y,type="n",xlab="Factor Score 1",ylab="Factor Score 2")
text(x,y,"1":"79",cex=0.75); title(main=quot,cex=0.8)

dev.off()
```

Output from R 3.2 The variances (squared standard deviations) are the five eigenvalues of the 5×5 correlation matrix R. Their sum is 5. The columns Components 1–5 of the matrix Loadings are the five (orthogonal) eigenvectors of R, normalized to 1. Eigenvalues and eigenvectors can also be obtained by the R-commands R<- cor(all5T) and

eigen(R)$values , eigen(R)$vectors

respectively.

```
"Principal components"
Importance of components:
                        Comp.1 Comp.2 Comp.3 Comp.4 Comp.5
Standard deviation       2.126 0.5162 0.3441 0.2608 0.1652
Proportion of Variance   0.904 0.0533 0.0237 0.0136 0.0055
Cumulative Proportion    0.904 0.9572 0.9809 0.9945 1.0000


Loadings:
      Comp.1 Comp.2 Comp.3 Comp.4 Comp.5
TpA    0.458  0.143 -0.019 -0.776  0.409
TpB    0.435 -0.709  0.056 -0.155 -0.530
TpH    0.442  0.513 -0.581  0.145 -0.427
TpK    0.444  0.349  0.779  0.258 -0.094
TpP    0.456 -0.304 -0.229  0.536  0.600
```

3.3 Auto-Correlation Function

The theoretical auto-covariance function $\gamma(h)$, $h = 1, 2, \ldots$ (see Appendix B.1) is estimated from the bivariate sample

$$(Y_1, Y_{h+1}), \ldots, (Y_{n-h}, Y_n) \tag{3.2}$$

of size $n - h$. On the basis of (3.2) the empirical auto-covariance $\hat{\gamma}(h) = c(h)$ is calculated by

$$c(h) = \frac{1}{n} \sum_{t=1}^{n-h} (Y_t - \bar{Y})(Y_{t+h} - \bar{Y}), \quad h = 0, 1, \ldots,$$

with the total mean $\bar{Y} = (1/n) \sum_{i=1}^{n} Y_i$.

Here, the estimator $\hat{\sigma}^2$ for the variance $\sigma^2 = \mathrm{Var}(Y_t)$, that is

$$\hat{\sigma}^2 = c(0) = \frac{1}{n} \sum_{t=1}^{n} (Y_t - \bar{Y})^2,$$

has the factor $1/n$ and not—as usual in standard statistics—the factor $1/(n-1)$.

The empirical auto-correlation $\hat{\rho}(h) = r(h)$ is gained by using $c(h)$, namely by

$$r(h) = \frac{c(h)}{c(0)} = \frac{\sum_{t=1}^{n-h} (Y_t - \bar{Y})(Y_{t+h} - \bar{Y})}{\sum_{t=1}^{n} (Y_t - \bar{Y})^2}, \quad h = 0, 1, \ldots.$$

We have $r(0) = 1$ and $|r(h)| \leq 1$. The quantities $r(h)$, plotted over $h = 1, 2, \ldots$, are also called the *correlogram* of the time-series. It is recommended to perform correlogram analysis in time series with $n \geq 50$ only, and to evaluate $r(h)$ only up to a *time lag* $h \leq [n/4]$; see Box & Jenkins (1976).

A test for a pure random series (or *white noise* process) is based on k values $r(h)$, $h = 1, \ldots, k$, of the correlogram. The hypothesis

H_0 : the time series is the realization of a white noise process

is rejected (level α) if, with the *Bonferroni*-bound $b_k = u_{1-\beta/2}/\sqrt{n}$, $\beta = \alpha/k$,

at least one of the values $|r(1)|, |r(2)|, \ldots, |r(k)|$ exceeds b_k.

The individual bound $b_1 = u_{1-\alpha/2}/\sqrt{n}$ is valid for a coefficient $r(h)$ with a time lag h specified in advance (for instance $h = 1$). Note that 5 % of the correlogram values of a pure random series (where H_0 is true!) exceeds—on the average—the individual bounds $\pm b_1$ (when $\alpha = 0.05$ was chosen).

Table 3.6 Auto-correlation function (correlogram) of temperature series up to lag 12—omitting
$h = 9, 10, 11$; for Hohenpeißenberg 1781–2010, Karlsruhe 1799–2008 and Potsdam 1893–2010

	Hohenpeißenberg			Karlsruhe			Potsdam		
h	$r_Y(h)$	$r_e(h)$	$r_W(h)$	$r_Y(h)$	$r_e(h)$	$r_W(h)$	$r_Y(h)$	$r_e(h)$	$r_W(h)$
1	0.295	0.115	0.076	0.329	0.110	0.112	0.355	0.146	0.123
2	0.254	0.068	0.055	0.312	0.101	−0.003	0.232	−0.009	0.000
3	0.175	−0.022	0.050	0.227	0.001	0.005	0.025	−0.254	−0.113
4	0.213	0.033	−0.053	0.211	−0.014	−0.060	0.071	−0.167	−0.148
5	0.117	−0.080	0.036	0.223	0.006	−0.003	0.121	−0.080	0.051
6	0.150	−0.038	0.077	0.233	0.036	0.104	0.128	−0.069	0.021
7	0.148	−0.036	0.031	0.240	0.058	0.095	0.206	0.037	0.207
8	0.111	−0.070	−0.015	0.181	−0.011	−0.043	0.173	0.004	−0.053
...	...	...	...	...	...	...	...	...	...
12	0.113	−0.030	−0.043	0.087	−0.079	−0.055	0.004	−0.125	−0.005
b_1	0.129	0.129	0.129	0.135	0.135	0.135	0.180	0.180	0.180
b_{12}	0.189	0.189	0.189	0.198	0.198	0.198	0.264	0.264	0.264

Y = yearly data, W = winter data, $e = Y - pol(4)$, the residuals of the yearly data from polynomial
trend (order 4). Individual (b_1) and simultaneous (b_{12}) bounds are added, $\alpha = 0.05$

Annual Temperature and Precipitation

The correlogram $r(h)$, $h = 1, \ldots, 12$, of the $n = 230$ annual *temperature* means
of Hohenpeißenberg has three values above the simultaneous bound b_{12} and most
values above b_1. These high (positive) auto-correlation values result from the trend
component of the series. If we remove a polynomial trend (of order 4) we have positive
and negative values, now all between the $\pm b_1$ bounds, see Table 3.6 and Fig. 3.5. The
same phenomenon (even more drastically) can be reported from the Karlsruhe series
(Table 3.6, but no plots). As already observed in Sect. 1.2, the winter temperature
series are much nearer to a pure random series than the annual temperature series
are. This is now confirmed by the auto-correlation functions reproduced in Table 3.6
and Fig. 3.6.

The Karlsruhe series of annual *precipitation* (without any trend removal)
is—according to its correlogram in Table 3.7—close to a pure random series; the
same is true for Potsdam (see also Fig. 3.8). But this is different from Hohenpeißenberg, where the precipitation series—as already mentioned above in 3.1—has more
auto-correlation structure; see the correlogram in Table 3.7 and Fig. 3.7.

| R 3.3 | Auto-correlation function of a time series by means of `acf`, but no plot
is specified within `acf` (plot=F). A needle-plot as in Fig. 3.5 is produced by the
user function `plotAcf`. Horizontal lines b_1 and b_{12} of individual and simultaneous
bounds are drawn.

```
attach(hohenTp)
quot<- "Hohenpeissenberg, Temperature, 1781-2010"; quot
```

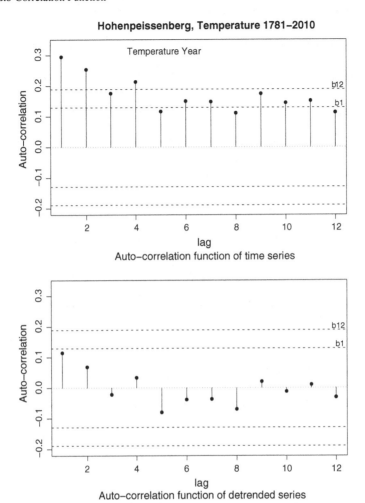

Fig. 3.5 Hohenpeißenberg, annual temperature means, 1781–2010. *Top* Auto-correlation function (correlogram) of the time series. *Bottom* Auto-correlation function (correlogram) of the residuals from polynomial trend (order 4). Individual (b_1) and simultaneous (b_{12}) bounds are drawn, $\alpha = 0.05$

```
plotAcf<- function(ACF,maxl,cylim,b1,bm){    #Needle-Plot of ACF
plot(1:maxl,ACF,pch=16,ylim=cylim,xlab="lag",
          ylab="Auto-correlation")
for (i in 1:maxl){
   segments(i,0.0,i,ACF[i])}                          #Needles
abline(h=0,lty=3); abline(h=c(-b1,-bm,b1,bm),lty=2)    #Bounds
text(maxl+0.1,b1+0.01,"b1",cex=0.7)
text(maxl+0.1,bm+0.01,"b",cex=0.7)
text(maxl+0.25,bm+0.01,maxl,cex=0.7)
```

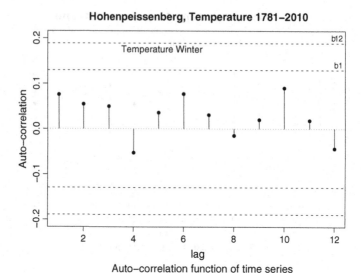

Fig. 3.6 Hohenpeißenberg, winter temperature means, 1781–2010. Auto-correlation function (correlogram) of the time series. Individual (b_1) and simultaneous (b_{12}) bounds are drawn, $\alpha = 0.05$

```
}

#Vector Y is the time series to be analyzed
Y<- Tyear/100; n<- length(Y);              maxl<- 12     #maximal lag

subtxt<- "Auto-correlation function of time series"; subtxt
zacf<- acf(Y,lag.max=maxl,type="corr",plot=F)              #no Plot

ACF<-zacf$acf[2:(maxl+1)]; ACF       #Output of $r(1)...r(maxl)

postscript(file="C:/CLIM/Acf.ps",height=20,width=12,horiz=F)
par(mfrow=c(2,1))

cylim<- c(-0.2,0.33)
b1<- qnorm(0.975)/sqrt(n);     bm<- qnorm(1-0.025/maxl)/sqrt(n)
plotAcf(ACF,maxl,cylim,b1,bm)               #Produce Needle-Plot
title(main=quot);   title(sub=subtxt,cex=0.7)

#---Analogously for the detrended series----------------------

dev.off()
```

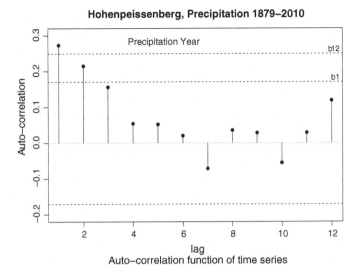

Fig. 3.7 Hohenpeißenberg, annual precipitation amounts, 1879–2010. Auto-correlation function (correlogram) of the time series. Individual (b_1) and simultaneous (b_{12}) bounds are drawn, $\alpha = 0.05$

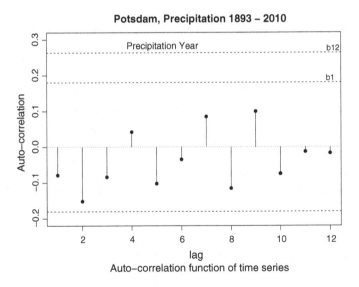

Fig. 3.8 Potsdam, annual precipitation amounts, 1893–2010. Auto-correlation function (correlogram) of the time series. Individual (b_1) and simultaneous (b_{12}) bounds are drawn, $\alpha = 0.05$

Table 3.7 Auto-correlation function (correlogram) of the annual precipitation amounts up to lag 12—omitting $h = 9, 10, 11$; for Hohenpeißenberg (H) 1879–2010, Karlsruhe (K) 1876–2008 and Potsdam (P) 1893–2010

h	1	2	3	4	5	6	7	8	...	12	b_1	b_{12}
H	0.273	0.215	0.156	0.05	0.05	0.02	−0.07	0.03	...	0.12	0.17	0.25
K	0.009	−0.116	−0.023	0.03	−0.02	−0.05	−0.07	0.08	...	0.03	0.17	0.25
P	−0.078	−0.151	−0.084	0.04	−0.10	−0.03	0.08	−0.12	...	−0.02	0.18	0.26

Individual (b_1) and simultaneous (b_{12}) bounds are added, $\alpha = 0.05$

Table 3.8 Conditional probabilities for exceeding threshold values (quantiles) Q_γ, for $\gamma = 0.50, 0.75, 0.90$

Conditional probability	Correlation $\rho = \rho_{X,Y}$							
	0.00	0.10	0.20	0.30	0.40	0.50	0.60	0.70
$\mathbb{P}\left(Y > Q_{0.50}^y \mid X > Q_{0.50}^x\right)$	0.50	0.53	0.56	0.60	0.63	0.67	0.70	0.75
$\mathbb{P}\left(Y > Q_{0.75}^y \mid X > Q_{0.75}^x\right)$	0.25	0.29	0.34	0.40	0.45	0.51	0.57	0.64
$\mathbb{P}\left(Y > Q_{0.90}^y \mid X > Q_{0.90}^x\right)$	0.10	0.14	0.17	0.24	0.29	0.39	0.45	0.53

Each entry is calculated by means of 40,000 simulations of a pair (X, Y) of two-dimensional Gaussian random variables

3.4 Prediction of Above-Average Values

Assume that we have calculated a certain value for the auto-correlation $r(1) = r(Y_t, Y_{t+1})$. Assume further, that we have just observed an above-average value of Y_t (or an extreme value of Y_t). What is the probability $\mathbb{P}$, that the next observation Y_{t+1} will be above-average (extreme), too?

To tackle this problem, let X and Y denote two random variables, with the coefficient $\rho = \rho_{X,Y}$ of the true correlation between them. We ask for the probability, that an observation X, being greater than a certain threshold value Q^x, is followed by an observation Y, exceeding a Q^y. If the X-value exceeds Q^x, then Table 3.8 gives (broken down according to the coefficient ρ) the probabilities $\mathbb{P}$ for the event, that the Y-value exceeds Q^y. As threshold values we choose *quantiles* Q_γ (also called $\gamma \cdot 100\%$ percentiles), for $\gamma = 0.5, 0.75, 0.90$. These threshold values could also be called: average value (more precisely an 50% value), upper 25% value, upper 10% value, respectively.

Examples:

1. Assume that X turns out to exceed $Q_{0.50}^x$ (X being an upper 50% value, shortly: being above-average). Then the probability that Y is above-average, too, equals

   50% for $\rho = 0$;   60% for $\rho = 0.30$;   70% for $\rho = 0.60$.
2. If X exceeds $Q_{0.90}^x$ (X being an upper 10% value), then the probability that Y is an upper 10 percent value, too, equals

   10% for $\rho = 0$;   24% for $\rho = 0.30$;   45% for $\rho = 0.60$.

In the sequel, X and Y will denote climate variables, where X is followed by Y.

Table 3.9 Hit ratios of the rules 1–6

Ex	$X \to Y$		$r(X,Y)$	$\mathbb{P}$		% $[Y \Diamond \bar{y} \mid X \Diamond \bar{x}]$[a]			
			Hohen			Berlin (%)	Hohen (%)	Brem (%)	Karls (%)
1	Tp Dec	Tp Jan	0.13	0.54	[> | >]	70	56	58	58
1	Tp Dec	Tp Feb	0.11	0.53	[> | >]	60	55	62	62
2	Tp Sep	Tp Oct	0.14	0.54	[> | >]	62	57	56	55
				0.54	[< | <]	62	52	48	54
3	Tp Nov	Tp May	0.04	0.51	[> | >]	50	52	52	55
	Pr Nov	Pr May	−0.02	0.50	[> | >]	50	43	42	42
4	Tp Aug	Tp Feb	0.08	0.53	[> | >]	73	52	62	56
	Pr Aug	Pr Feb	0.04	0.51	[> | >]	50	47	40	50
5	Tp Sum	Tp Win	0.06	0.51	[> | >]	–	53	62	54
				0.49	[< | >]	–	47	38	46
	Pr Sum	Pr Win	−0.02	0.50	[> | >]	–	42	40	41
				0.50	[< | >]	–	58	60	59
6	Tp Win	Tp Sum	0.16	0.55	[> | >]	–	55	56	51
				0.55	[< | <]	–	62	52	55
				0.45	[> | <]	–	38	48	45
	Pr Win	Pr Sum	0.22	0.57	[> | >]	–	62	48	50
				0.43	[> | <]	–	42	52	43

Explanations in the text

[a] $\Diamond$ stands for a ">" or a "<" sign

Application to Climate Data

Once again, only the results for the adjusted series, that are the figures in Tables 3.1, 3.2 and 3.3 not in parenthesis, are discussed.

The absolute value of most auto-correlations $r(1)$ falls into the interval from 0.0 to 0.2. Hence the ratio of hits—when observing an above-average climate value and predicting the same for the next observation—lies between 50 and 56 %, according to Table 3.8. (This is to compare with 50 %, when predicting 'above average' independently of the present observation.) These modest chances of a successful prediction will find their empirical counterparts in Table 3.9.

The daily temperatures, with $r(1) > 0.70$, have a hit ratio >75 % for the prediction *above-average* → *above-average*. If we have an upper 10 % day, then we can predict the same for the next day with success probability above 53 % (to compare with 10 %).

The auto-correlation coefficients $r(1)$ of the daily precipitation amounts, listed in Tables 3.1, 3.2 and 3.3, are not very meaningful, since the half of all days is without any precipitation. If we introduce the dichotomy (Prec > 0 or Prec = 0, if there is or there is no precipitation), we obtain for the 365*4 days of the years 2004–2007 at Hohenpeißenberg and in Karlsruhe the following 2×2 frequency tables.

Hohenpeißenberg present day	Succeeding day Prec = 0	Prec > 0	Σ
Prec = 0	493	225	718
	0.687	0.313	1.0
Prec > 0	225	516	741
	0.304	0.696	1.0
Σ	718	741	1459
	0.492	0.508	1.0

Karlsruhe present day	Succeeding day Prec = 0	Prec > 0	Σ
Prec = 0	536	243	779
	0.688	0.312	1.0
Prec > 0	243	437	680
	0.357	0.643	1.0
Σ	779	680	1459
	0.534	0.466	1.0

Evaluating the relative frequencies of the tables, we can report the following results (simplifying "precipitation" to "rain").

If a day keeps dry, we can say the same for the next day with a hit ratio of $\approx 69\%$ [Hohenpeißenberg and Karlsruhe].

If a day is rainy, we can predict rain for the following day with hit ratios of $\approx 70\%$ [Hohenpeißenberg] resp. $\approx 64\%$[Karlsruhe].

That is, we have the percentages

prediction dry → dry: $\approx 69\%$ [Hohenpeißenberg] $\approx 69\%$ [Karlsruhe]

prediction rainy → rainy: $\approx 70\%$ [Hohenpeißenberg] $\approx 64\%$ [Karlsruhe].

When evaluating "weather rules" concerning temperature and precipitation, numerical schemes of the type of Tables 3.1, 3.2 and 3.3 or of the above 2 × 2 tables become important.

The analysis of daily precipitation data is continued in Sects. 6.3, 6.4 and 6.5.

3.5 Folk Sayings

Folk (or country) sayings about weather relate to

a particular region (presumably not covered here)

a particular time epoch (here centuries are involved)

and to the crop (Malberg 2003). The former weather observers (from the country or from

monasteries) without modern measuring, recording, and evaluation equipments were pioneers of weather forecasting.

The following sayings are selected from Malberg (2003) and from popular sources. We kept the German language, but we have transcribed them in Table 3.9.

Persistence rules

Ex. 1: Ist Dezember lind → der ganze Winter ein Kind

Ex. 2: Kühler September → kalter Oktober

Six-months rules

Ex. 3: Der Mai kommt gezogen ← wie der November verflogen

Ex. 4: Wie der August war → wird der künftige Februar

Yearly-balance rules

Ex. 5: Wenn der Sommer warm ist → so der Winter kalt

Ex. 6: Wenn der Winter kalt ist → so der Sommer warm

The columns of Table 3.9 present

- transcription of the weather rules 1–6, with Tp standing for temperature and Pr for precipitation
- correlation coefficient $r = r(X, Y)$ from the Hohenpeißenberg data
- conditional probability $\mathbb{P}(Y > Q_{0.5}^y | X > Q_{0.5}^x)$, belonging to the r-value according to Table 3.8
- percentage $\% [Y > \bar{y} | X > \bar{x}]$ of cases with an above-average X-value, in which an above-average Y-value follows. This is given for Berlin-Dahlem 1908–1987 (Malberg 2003), Hohenpeißenberg, Bremen, Karlsruhe.

Rule 2 aims at the percentage $\% [Y < \bar{y} | X < \bar{x}]$, rule 5 at $\% [Y < \bar{y} | X > \bar{x}]$, rule 6 at $\% [Y > \bar{y} | X < \bar{x}]$. These percentages are presented, too, in addition to the percentage $\% [Y > \bar{y} | X > \bar{x}]$.

The hit ratios, gained from the Hohenpeißenberg and from the Karlsruhe data, are rather poor and cannot confirm the rules (Bremen performs only slightly better). At most the persistence rules find a weak confirmation. In some cases another version of the rule (Ex. 2) or even the opposite rule (Ex. 5, Ex. 6) are proposed by our data.

With one or two exceptions the Berlin-Dahlem series brings higher hit ratios than the series from Hohenpeißenberg or Bremen, Karlsruhe. The reason could be, that the Dahlem series is shorter and is perhaps (climatically) nearer to the place of origin of the rules.

Note that the theoretical $\mathbb{P}$ values from Table 3.8, given here for the r-values of the station Hohenpeißenberg, are consistent with the empirical percentages in Table 3.9, evaluated for temperature at the station Hohenpeißenberg.

| R 3.4 | Calculating the hit ratios for the weather rules 1 and 2, by applying the user function CondFrequ. For the n-dimensional vector $x = (x(1), \ldots, x(n))$, x[condition] selects the x-components $x(i)$ for those cases i, where condition is fulfilled. So x0 contains the x-values for cases, where x is above-average, yx0 the y-values for cases, where y *and* x are above-average.

```
attach(hohenTp)
#Number of cases with above-/below-average values
CondFrequ<- function(x,y){      #Conditional frequencies y|x
x0<- x[x > mean(x)]
yx0<- y[x > mean(x) & y > mean(y)]
x1<- x[x < mean(x)]
```

```
yx1<- y[x < mean(x) & y < mean(y)]
c("x>"=length(x0),"x>&y>"=length(yx0),
                "y>|x>"= length(yx0)/length(x0),
  "x<"=length(x1),"x<&y<"=length(yx1),
                "y<|x<"=length(yx1)/length(x1)) }
#------------------------------------------------------------
"x=Dec -> y=Jan"
x<- dcly/10; y <- jan/10
c("mean(x)"=mean(x),"mean(y)"=mean(y),"cor(x,y)"=cor(x,y))
CondFrequ(x,y)


"x=Sep -> y=Oct"
x<- sep/10;  y <- oct/10
c("mean(x)"=mean(x),"mean(y)"=mean(y),"cor(x,y)"=cor(x,y))
CondFrequ(x,y)
```

| Output from R 3.4 | $y > |x >$ resp. $y < |x <$ denotes the relative number of cases with above-average values resp. below-average values, followed by cases of the same kind.

Example: We have $63/112 = 0.5625$.

```
   "x=Dec -> y=Jan"
 mean(x)   mean(y)  cor(x,y)
 -0.9461   -1.9639    0.1316
      x>      x>&y>      y>|x>         x<      x<&y<      y<|x<
  112.00      63.00     0.5625     118.00      58.00     0.4915

   "x=Sep -> y=Oct"
 mean(x)   mean(y)  cor(x,y)
 11.6557    7.1461    0.1391
      x>      x>&y>      y>|x>         x<      x<&y<      y<|x<
  120.00      68.00     0.5667     110.00      57.00     0.5182
```

Chapter 4
Model and Prediction: Yearly Data

In the following we discuss statistical models, which are supposed (i) to describe the mechanism how a climate series evolves, and which can support (ii) the prediction of climate values in the next year(s). Time series models of the ARMA-type, as described in the Appendix B.3, will stand in the center of our analysis. These models are applied to the series of differences of consecutive time series values; this "differenced" series is considered as sufficiently "trendfree".

Predictions are calculated as real forecasts: The prediction for the time point t is based on information up to time $t - 1$ only. Residuals from the predictions are formed and analyzed by means of auto-correlation functions and by GARCH-models. The sum of squared residuals serves as a goodness-of-fit measure. On the basis of this measure, the ARMA-models are compared with (left-sided) moving averages. Finally, the annual precipitation series are investigated by means of GARCH-models.

4.1 Differences, Prediction, Summation

Let Y be the time series of N yearly climate records; that is, we have the data $Y(t)$, $t = 1, \ldots, N$. In connection with time series models and prediction, the trend of the series is preferably removed by forming differences of consecutive time series values. From the series Y we thus arrive at the *differenced* series X, with

$$X(t) = Y(t) - Y(t - 1), \quad t = 2, \ldots, N, \quad [X(1) = 0]. \tag{4.1}$$

Table 4.1 shows that the yearly changes X of temperature have mean ≈ 0 and an average deviation (from the mean 0) of ≈ 1 (°C), at all four stations. The first order auto-correlations $r(1)$ of the differences X lie in the range $-0.4 \ldots -0.5$. After an increase of temperature follows—by tendency—an immediate decrease in the next year, and vice versa; see also the upper plots of Figs. 4.1, 4.2.

H. Pruscha, *Statistical Analysis of Climate Series*,
DOI: 10.1007/978-3-642-32084-2_4, © Springer-Verlag Berlin Heidelberg 2013

Table 4.1 Differences X of temperature means (°C) in consecutive years

Station	N	Mean	Standard deviation	$r(1)$	$r(2)$	$r(3)$
Bremen	121	0.003	0.856	−0.384	0.060	−0.162
Hohenpeißenberg	230	−0.004	1.002	−0.465	0.021	−0.076
Karlsruhe	210	0.011	0.921	−0.489	0.057	−0.052
Potsdam	118	0.002	0.926	−0.402	0.074	−0.200

Mean value, standard deviation and the first 3 auto-correlation coefficients of X are given

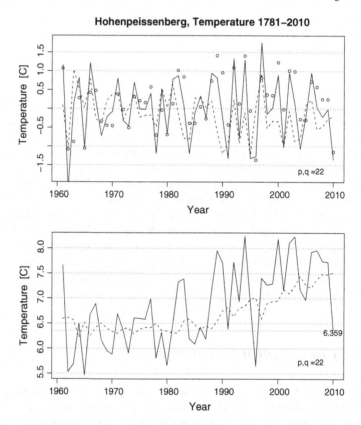

Fig. 4.1 Hohenpeißenberg, annual temperature means, 1781–2010. *Top* Differenced time series, with ARMA-predictions (*dashed line*) and with residual values (as *circles o*). *Bottom* Time series of annual temperature means (°C), together with the ARIMA-prediction (*dashed line*). The last 50 years are shown

We now consider the differenced time series $X(t)$ as sufficiently "trendfree" and try to fit an ARMA(p,q)-model. Such a model obeys the equation

$$X(t) = \alpha_p X(t-p) + \cdots + \alpha_2 X(t-2) + \alpha_1 X(t-1)$$
$$+ \beta_q e(t-q) + \cdots + \beta_2 e(t-2) + \beta_1 e(t-1) + e(t), \qquad (4.2)$$

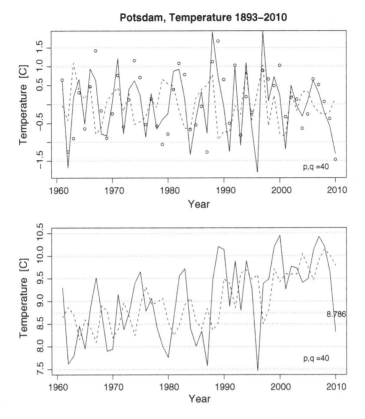

Fig. 4.2 Potsdam, annual temperature means, 1893–2010. Legend as in Fig. 4.1

with error (residual) variables $e(t)$. For each time point t we can calculate a prognosis $\hat{X}(t)$ for the next observation $X(t)$, called ARMA-prediction. This is done on the basis of the preceding observations $X(t-1)$, $X(t-2)$, ... The prognosis $\hat{X}(t)$ is computed as in Eq. (4.2), but setting $e(t)$ zero, while the other variables $e(t-1)$, $e(t-2)$, ... are recursively gained as described around Eq. (B.13) in the Appendix. We have then the ARMA-prediction

$$\hat{X}(t) = \alpha_p X(t-p) + \cdots + \alpha_2 X(t-2) + \alpha_1 X(t-1)$$
$$+ \beta_q e(t-q) + \cdots + \beta_2 e(t-2) + \beta_1 e(t-1). \tag{4.3}$$

Equation (4.3) constitutes the Box & Jenkins forecast formula for *time lead $l = 1$*, see Eq. (B.18) (setting there $T = t - 1$; l-steps forecasts follow in Sect. 8.2).

From the differenced series X we get back the original series Y by recursive summation (also called integration): $Y(t) = Y(t-1) + X(t)$. The prediction $\hat{Y}(t)$ for $Y(t)$ is gained by

$$\hat{Y}(t) = Y(t-1) + \hat{X}(t), \quad t = 2, \ldots, N; \qquad \hat{Y}(1) = Y(1).$$

Note that the residuals fulfill the equation

$$X(t) - \hat{X}(t) = Y(t) - \hat{Y}(t).$$

This procedure is called the ARIMA-method, the variables $\hat{Y}(t)$ are referred to as ARIMA-predictions for $Y(t)$.

Updating, Goodness-of-Fit

The calculation of $\hat{X}(t)$ has to be based on the information up to time $t-1$, so that we have to demand the same for the estimates of the coefficients α_i, β_j appearing in Eq. (4.3). For this reason, we estimate α_i, β_j for each time point t (greater than a starting value t_0) anew, namely on the basis of the data

$$X(1), \ldots, X(t-1), \quad t \geq t_0 + 1. \tag{4.4}$$

For the estimation procedure we need a minimum sample size t_0; thus we have the $\hat{X}(t)$ at our disposal only from t_0 onwards. However, the estimation of the α and β uses the series from the beginning (time point 1) upwards. The coefficients estimated on the basis of (4.4) could be denoted by $\alpha_i^{[1,t-1]}$, $\beta_j^{[1,t-1]}$ instead of α_i, β_j (we write α, β for the unknown coefficients as well as for their estimates). The goodness of the prediction and hence the goodness-of-fit of the ARMA-model is assessed by the principle of *residual-sum-of-squares*. More detailed, we build the mean value of the squared residuals (MSQ) and extract then the square root, that is

$$RootMSQ = \sqrt{(1/N_0) \cdot \sum\nolimits_{t=t_0+1}^{N} \left(X(t) - \hat{X}(t)\right)^2}, \quad N_0 = N - t_0. \tag{4.5}$$

The smaller the *RootMSQ* value, the better is the fit of the model. Due to $X(t) - \hat{X}(t) = Y(t) - \hat{Y}(t)$, the prediction $\hat{Y}(t)$ for $Y(t)$ is as good as the prediction $\hat{X}(t)$ for $X(t)$, namely by (4.5)

$$RootMSQ = \sqrt{(1/N_0) \cdot \sum\nolimits_{t=t_0+1}^{N} \left(Y(t) - \hat{Y}(t)\right)^2}. \tag{4.6}$$

Finally, we can build the standardized *RootMSQ* measure rsq, i.e.,

$$rsq = \frac{RootMSQ}{sd(X)}, \tag{4.7}$$

with the standard deviation sd(X) of the N_0 values of the differenced series $X(t)$, $t = t_0 + 1, \ldots, N$.

Table 4.2 ARIMA-method for the annual temperature means (°C)

sd	Order p, q	ARMA-coefficients α_i	β_j	Root MSQ	ARIMA-prediction 2008–2010	2011
B	3, 1	0.228, 0.076, −0.193	−0.863	0.805	10.11, 9.96, 9.85	9.33
0.897	obs:				10.10, 9.98, 8.34	10.14
H	2, 2	−0.639, 0.105	−0.18, −0.67	0.762	7.49, 7.48, 7.52	7.26
0.981	obs:				7.74, 7.72, 6.38	8.48
P	4, 0	−0.56, −0.31, −0.41, −0.27		0.869	10.15, 10.01, 9.79	9.27
0.963	obs:				10.22, 9.63, 8.32	10.14
					2006–2008	2009
K	3, 1	0.001, −0.02, −0.12	−0.809	0.687	11.32, 11.47, 11.53	11.49
0.844	obs:				11.61, 11.84, 11.59	

Coefficients (calculated from the whole series), goodness-of-fit, predictions for the years 2008–2011 (Karlsruhe: 2006–2009), with actually observed values beneath. Further: sd = sd(X) denotes the standard deviation of the $N_0 = N - t_0$ values of the differenced series $X(t)$, $t = t_0 + 1, \ldots, N$. B Bremen, H Hohenpeißenberg, K Karlsruhe, P Potsdam

4.2 ARIMA Method for Yearly Temperature Means

Now $Y(t)$ denotes the temperature mean of the year t and $X(t)$—according to Eq. (4.1)—the differenced series, i.e., the series of the yearly changes, see Figs. 4.1 and 4.2 (upper plots). It is the time series X to which an ARMA(p,q)-model is applied.

We choose (for the prediction) the starting year $t_0 = [N/5]$ and order numbers (p, q) as small as possible, such that an increase of these numbers brings no essential improvement of the goodness measure *RootMSQ*. For the Hohenpeißenberg data we get $p = q = 2$ and therefore the ARMA(2,2)-model

$$X(t) = \alpha_2 X(t - 2) + \alpha_1 X(t - 1) + \beta_2 e(t - 2) + \beta_1 e(t - 1) + e(t) \qquad (4.8)$$

(for Bremen and Karlsruhe we obtain $p = 3, q = 1$, and for Potsdam $p = 4, q = 0$). Table 4.2 shows the estimated coefficients $\alpha_i = \alpha_i^{[1,N]}$ and $\beta_j = \beta_j^{[1,N]}$ for the whole series (used to produce the forecast for the year 2011; Karlsruhe: 2009). As a rule, at least one α and one β are significantly different from zero. Further, the table offers the forecasts for the three years 2008–2010 as well as for the year 2011, each time on the basis of the preceding years. For Karlsruhe, we have predictions for the years 2006–2009 instead of 2008–2011. The prognoses for 2008 are quite good (exception Hohenpb); the relatively low temperatures of the year 2010 are overestimated by our prediction, compare also Figs. 4.1 and 4.2 (lower plots). These plots also show the smoothing character of the ARIMA-predictions. For a clearer presentation we confine ourselves to the reproduction of the last 50 years (but for calculating the coefficients α, β, the series was used from its beginning, of course). In contrast to the overestimated 2010-values, the comparatively high temperature means of the year 2011 are underestimated.

We derive the standardized *RootMSQ* measures (4.7) for annual temperature from Table 4.2 and obtain the following rsq-values, which are better for Hohenpeißenberg and Karlsruhe than for Bremen and Potsdam:

Bremen 0.897,   Hohenpb. 0.777,   Karlsruhe 0.814,   Potsdam 0.903

Remark A: We repeat that the *RootMSQ*-values 0.805, 0.762, 0.869, 0.687 for the four stations were calculated with predictions $\hat{X}(t)$, gained from Eq. (4.3) with coefficients $\alpha_i^{[1,t-1]}$, $\beta_j^{[1,t-1]}$. This was described above around (4.4). Let us call this procedure the *forecast approach* in regression analysis. In *standard regression* analysis the coefficients $\alpha_i = \alpha_i^{[1,N]}$, $\beta_j = \beta_j^{[1,N]}$ are computed only once for the whole series and are used for each $\hat{X}(t)$, $t = 1, \ldots, N$, according to (4.3). Here, in the standard approach, we could take a t_0-value as small as $t_0 = \max(p, q)$. For the sake of comparability however, we choose once again $t_0 = [N/5]$ as starting point for the predictions. We arrive at the following *RootMSQ*-values

	Bremen	Hohenpeißenberg	Karlsruhe	Potsdam
(p, q) *RootMSQ*	(3,1) 0.729	(2,2) 0.732	(3,1) 0.658	(4,0) 0.807

within standard regression, all four values being smaller than those of Table 4.2: The ARMA-model fits better, when combined with the standard approach of regression analysis. But that approach violates the forecast principle, advocated (e.g., around (4.4)) and applied throughout this text. To repeat, this principle seems to be more appropriate for (yearly updated) climate series.

R 4.1 Yearly temperature data: Differencing, ARMA-model for the differenced series, prediction for the differenced series and for the integrated series (=ARIMA-prediction), residual analysis. Note that the ARIMA-residuals are identical with the residuals of the detrended series, which can be checked by the supplementary program part. The forecast regression method is realized by the user function `armat`. Herein, for each $t = t_0, \ldots, N$, the R function `arma` operates on the data vector $Y(s)$, $s = 1, \ldots, t$, and `parma[t+1]` contains the prediction on the basis of $Y[1:t]$.

```
attach(bremenTp)
quot<- "Bremen, Temperature, 1890-2010"; quot

library(TSA)                         #see Cran-Software-Packages
#-------------------------------------------------------------------
armat<- function(Y,n,tst,ma,mb){     #forecast regression approach
#parmat vect of dim n+1,components 1,..,tst filled with mean val
parmat<- 1:(n+1); parmat[1:tst]<- mean(Y[1:tst])

for(t in tst:n) {
art<- arma(Y[1:t],order=c(ma,mb))
coef<- art$coef;   resi<- art$residuals
parma<- coef[ma+mb+1]                        #intercept theta
a<- rep(0,times=12); b<- a; mc<- pmax(ma,mb)
if (ma > 0) for (m in 1:ma){a[m]<- coef[m]}   #alpha-coeff.
```

```
if (mb > 0) for (m in 1:mb){b[m]<- coef[ma+m]}        #beta-coeff.
for (m in 1:mc){parma<- parma + a[m]*Y[t+1-m] + b[m]*resi[t+1-m]}
parmat[t+1]<- parma
}
return(parmat)  }                      #return prediction vector parmat


#--------------Data preparation, Differencing------------------
mon12<- data.frame(bremenTp[,3:14])/10;           #select jan-dec
Yje<- rowMeans(mon12)         #more precise than Yje<- Tyear/100
N<- length(Year);                 Dy<- Yje
Dy[1]<- 0; Dy[2:N]<- Yje[2:N]-Yje[1:(N-1)] #Dy differenced series

ma<- 3; mb<- 1; tst<- trunc(N/5);  ts1<- tst+1
c("ArOrder"=ma,"MAOrder"=mb,"Start"=tst,"Ncut"=N-tst,
            "StdevDYcut"=sqrt(var(Dy[ts1:N])))

# ---------- ARIMA(p,q)-model and ARIMA-prediction----- ---------
#  a) Differenced series Dy--------------------
darma<- arma(Dy,order=c(ma,mb));  "Results for whole diff series"
summary(darma)                          #for output only
"Forecast regression approach, differenced series"
parmat<- armat(Dy,N,tst,ma,mb)              #vector i=1:(N+1)
pred<- parmat[ts1:N]; dres<- Dy[ts1:N]-pred #vector i=1:(N-(tst))
msq<- mean(dres*dres)
c("MeanDres"=mean(dres),"StdDres"=sqrt(var(dres)),
   "MSQ (differenced)"=msq,"RootMSQ"=sqrt(msq))

"Auto-correlations of residuals, differenced series"
acf(dres,lag.max=8,type="corr",plot=F)

#  b) Integrated series YIarma--------------------
YIarma<- Yje; YIarma[2:N]<- Yje[1:(N-1)]+parmat[2:N]  #vect i=1:N
"Observations and ARIMA-predictions for last decade"
Yje[(N-9):N]; YIarma[(N-9):(N)]
c("Forecast NewYear"= Yje[N],parmat[N+1],Yje[N]+parmat[N+1])
```

Supplement

```
"Test: ARIMA-residuals = ARMA-residuals of differenced series"
"Test: same MSQ and auto-corr values as above"
YIres<- Yje - YIarma                 #YIres[ts1..] = dres[1...]
YIrer<- YIres[ts1:N];  misq<- mean(YIrer*YIrer)
c("MSQ (integrated)"=misq,"RootMSQ"=sqrt(misq))
"Auto-correlations of ARIMA-residuals"
acf(YIrer,lag.max=8,type="corr",plot=F)
```

Output from R 4.1 ARMA-output for differenced series, ARIMA-prediction for the integrated series. Residuals with *RootMSQ*-value and auto-correlation function. Bremen, Temperature 1890–2010.

```
"Results for whole diff series"
Model: ARMA(3,1),   Coefficient(s):
            Estimate   Std. Error   t value Pr(>|t|)
ar1          0.22830     0.10372      2.20    0.028 *
ar2          0.07624     0.09927      0.77    0.442
ar3         -0.19291     0.09659     -2.00    0.046 *
ma1         -0.86357     0.05999    -14.39    <2e-16 ***
intercept    0.01167     0.00969      1.20    0.229
```

```
"Forecast regression approach, differenced series"
        MeanDres        StdDres       MSQ (differenced)      RootMSQ
       -0.07175        0.80571           0.64763            0.80475
```

```
"Auto-correlations of residuals, differenced series"
    1      2       3      4       5       6       7       8
 0.114  0.049 -0.122 -0.097 -0.025 -0.091 -0.020 -0.058
```

```
"Observations and ARIMA-predictions for last decade"
9.400 9.900 9.5333 9.625 9.658 10.192 10.542 10.100 9.983 8.342
9.856 9.424 9.5494 9.687 9.594  9.693  9.887 10.114 9.960 9.852
```

```
Forecast NewYear      8.34167      +      0.99195      =      9.33362
```

Comparison with Moving Averages

Alternatively, prediction according to the method of left-sided moving averages can be chosen. As prediction $\hat{Y}(t)$ (for $Y(t)$ at time point t) we take the average of the preceding observations $Y(t-1)$, $Y(t-2)$, ..., $Y(t-k)$. The number k of the "depth" of averaging denotes the number of lagged variables and hence the number of years involved in the average. For the reason of comparability, we choose here a starting point $t_0 = [N/5]$, too. Once again by Eq. (4.6) we calculate the goodness of this prediction method. Table 4.3 demonstrates, that for a depth k smaller than 5 (Po.), 6 (Ho.), 7 (Br.,Ka.) the *RootMSQ*-values of the ARIMA-method are not improved. Notice that the latter method only needed $p + q = 4$ lagged variables. In the case of Potsdam, the autoregressive model of order $p = 4$ performs only little better than the—closely related—(left-sided) moving averages with $k = 4$.

4.3 ARIMA-Residuals: Auto-Correlation, GARCH Model

Having calculated the ARIMA-predictions $\hat{Y}(t)$ for $Y(t)$, $t = t_0 + 1, \ldots, N$, we then build residuals

$$e(t) = Y(t) - \hat{Y}(t), \quad t = t_0 + 1, \ldots, N, \tag{4.9}$$

Table 4.3 Left-sided moving averages for annual temperature means

Depth	*RootMSQ*			
k	Bremen	Hohenpeißenberg	Karlsruhe	Potsdam
4	0.839	0.794	0.722	0.885
5	0.830	0.780	0.708	0.865
6	0.818	0.761	0.695	0.851
7	0.800	0.756	0.679	0.832
8	0.794	0.753	0.678	0.819
ARIMA (p, q)	(3, 1) 0.805	(2, 2) 0.762	(3, 1) 0.687	(4, 0) 0.869

Depth k of averaging and the resulting goodness-of-fit *RootMSQ* are listed. The latter is given for the ARIMA-method, too (see Table 4.2)

Table 4.4 Auto-correlation function $r_e(h)$, up to time lag $h = 8$ (years), of the ARIMA-residuals $e(t)$, together with individual bounds b_1 and simultaneous bounds b_8 [level 0.05]

	$r_e(1)$	$r_e(2)$	$r_e(3)$	$r_e(4)$	$r_e(5)$	$r_e(6)$	$r_e(7)$	$r_e(8)$	b_1	b_8
B	0.114	0.049	−0.122	−0.10	−0.02	−0.09	−0.02	−0.06	0.199	0.278
H	−0.123	0.103	−0.118	0.02	−0.03	0.05	0.02	−0.02	0.144	0.201
K	0.136	0.027	−0.089	−0.17	−0.10	−0.01	−0.01	−0.03	0.151	0.211
P	0.102	−0.039	−0.149	−0.20	−0.23	−0.09	0.02	0.06	0.201	0.281

Annual temperature means at the stations *B* Bremen, *H* Hohenpeißenberg, *K* Karlsruhe, *P* Potsdam

from these predictions; see Figs. 4.1 and 4.2 (upper plots). Note that we already used these residuals in Eq. (4.6); as stated above we also have $e(t) = X(t) - \hat{X}(t)$. We ask now for the structure of the residual time series $e(t)$, $t = t_0 + 1, \ldots, N$. The values of the auto-correlation function $r_e(h)$, $h = 1, \ldots, 8$, are close to zero, cf. Table 4.4. The bound for the maximum of $|r_e(h)|$, $h = 1, \ldots, 8$ (i.e., the simultaneous bound with respect to the hypothesis of a pure random series), already used in 3.3, equals

$$b_8 = u_{1-0.025/8}/\sqrt{N_0} \quad \text{[significance level 0.05, } N_0 = N - t_0],$$

and is not exceeded; even the bound $b_1 = u_{0.975}/\sqrt{N_0}$ for an individual $|r_e(h)|$ is exceeded only one times ($r_e(5)$ for Potsdam). We can assume, that the series $e(t)$ consists of uncorrelated variables, for each of the four stations.

Next we ask, whether the (true) variances of the ARIMA-residuals $e(t)$ are constant over time—or whether periods of stronger and periods of minor oscillation alternate. To this end, we calculate—moving in 5-years time blocks $[t - 4, t]$—the empirical variances $\hat{\sigma}^2(t)$ of the $e(t - 4), \ldots, e(t)$. The roots $\hat{\sigma}(t)$, plotted in Fig. 4.3, form an oscillating line around the value 0.76 (see the *RootMSQ*-value for Hohenpeißenberg in Table 4.2), but a definite answer to the above question can not be given. A possibly varying oscillation of the series $e(t)$ may be explained by a GARCH-structure, which we are going to define next.

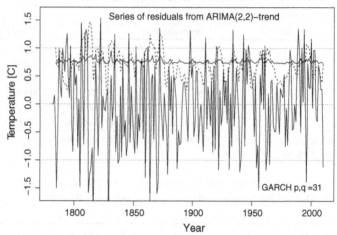

Fig. 4.3 Hohenpeißenberg, annual temperature means, 1781–2010. Time series of ARIMA-residuals (*zigzag line*), standard deviation $\hat{\sigma}$ of left-sided moving (5-years) blocks (*dashed line*), GARCH-predictions for σ (*solid line* around 0.76)

GARCH-Modeling the Residuals

A zero-mean process of uncorrelated variables $Z(t)$ is called a GARCH(p,q)-process ($p, q \geq 0$), if the (conditional) variance $\sigma^2(t)$ of $Z(t)$, given the information up to time $t-1$, fulfills the ARMA(p,q)-type equation

$$\sigma^2(t) = \alpha_p Z^2(t-p) + \cdots + \alpha_2 Z^2(t-2) + \alpha_1 Z^2(t-1) + \alpha_0$$
$$+ \beta_q \sigma^2(t-q) + \cdots + \beta_1 \sigma^2(t-1), \quad t = 1, 2, \ldots, \qquad (4.10)$$

(α's, β's nonnegative; see Kreiß and Neuhaus (2006); Cryer and Chan (2008)).

The GARCH-process $Z(t)$ can iteratively be generated by the equation

$$Z(t) = \sigma(t) \cdot \epsilon(t), \quad t = 1, 2, \ldots,$$

where $\sigma(t)$ obeys Eq. (4.10) and where $\epsilon(t)$ is a pure (0,1)-random series (independently distributed).

Order numbers (p, q) are to be determined (here $p = 3$, $q = 1$) and $p + q + 1$ coefficients α, β must be estimated. Then we build predictions $\hat{\sigma}^2(t)$ for the series $\sigma^2(t)$ in this way: Let the time point t be fixed. Having observed the preceding $Z(t-1)$, $Z(t-2)$, ..., and having already computed $\hat{\sigma}^2(t-1)$, $\hat{\sigma}^2(t-2)$, ..., then we put $\hat{\sigma}^2(t)$ according to Eq. (4.10), but with $\sigma^2(t-s)$ replaced by $\hat{\sigma}^2(t-s)$. Here the first q $\hat{\sigma}^2$-values must be predefined, for instance by the empirical variance of the time series Z. We are calling $\hat{\sigma}^2(t)$ the GARCH-prediction for the variance $\sigma^2(t)$.

Now we apply this method to our data and put $Z(t) = e(t)$, the (uncorrelated) ARIMA-residuals from Eq. (4.9). For the Hohenpeißenberg series we estimate the coefficients α_i and β_1 of the GARCH(3,1)-model, and calculate the GARCH-predictions $\hat{\sigma}^2(t)$, see Fig. 4.3. By means of the GARCH-residuals

$$\hat{\epsilon}(t) = e(t)/\hat{\sigma}(t)$$

we check the adequacy of the model: The mean and variance of $\hat{\epsilon}(t)$ are ≈ 0 and 1, resp., and the auto-correlation function $r_{\hat{\epsilon}}(h)$ runs near along the zero line, namely with values $0.013, 0.007, -0.014, 0.028, \ldots, -0.057$, for $h = 1, 2, 3, 4, \ldots, 8$.

Thus, the GARCH-model fits well to our series $e(t)$ of ARIMA-residuals.

The coefficients, the standard error of their estimation, and their quotient, i.e., the t-test statistic, are

	α_0	α_1	α_2	α_3	β_1
Coefficient	0.495	0.0260	0.0703	0.000	0.060
Standard error	10.7	0.107	0.580	1.50	20.3
t-test		0.243	0.121	0.000	0.003

With very small t-values and corresponding P-values near 1, the value zero for the coefficients α_1, α_2, α_3, β_1 is plausible, and thus a constant $\sigma(t)$-series can be assumed. Accordingly, the GARCH-predictions for $\sigma(t)$ reproduce in essence the horizontal line 0.76, see Fig. 4.3 and the *RootMSQ*-value in Table 4.2. This means that we can consider $e(t)$ as a series of uncorrelated variables with (conditionally and unconditionally) *constant* variance $\sigma^2(t) = \sigma^2$, i.e., as a $(0, \sigma^2)$-white noise process. From there we can state, that the differenced sequence $X(t)$ can sufficiently well be fitted by an ARMA-model, since the latter demands a white noise error process.

R 4.2 GARCH-model for the time series of ARIMA-residuals, that are the residuals from the ARIMA(2,2)-trend acc. to Sect. 4.2 (stored in the file HoT22Res.txt, with two variables Year, Y). Estimations for the coefficients of the GARCH(3,1)-model by the R-function garch, GARCH-prediction for $\sigma(t)$ twice, one time per R-function fitted.values, one time –in the supplement– per user function garchpr with identical results. Further: empirical estimation of $\sigma(t)$ by means of calculating the standard deviation in moving time blocks. The plot produces Fig. 4.3.

```
library(TSA)                        #see Cran-software-packages
Htpres<- read.table("C:/CLIM/HoT22Res.txt",header=T)
attach(Htpres)
quot<- "Hohenpberg, Temp 1781-2010, Residuals from trend"; quot
N<- length(Year); sde<- sqrt(var(Y))
c("Mean Y"=mean(Y),"Stdev Y"=sde,"Number Years"=N)

#---------R function garch----------------------------------
ma<- 3;  mb<- 1;  mc<- pmax(ma,mb)+1    #GARCH-order ma,mb <= 4
```

```
c("ma"=ma, "mb"=mb, "mc"=mc)
ord<- c(mb,ma)                                  #reversed order input
zgarch<- garch(Y,order=ord,maxiter=40)
summary(zgarch)                                 #a.o. diagnostic tests

 #---------GARCH-prediction, sigma(t) estimation---------------
"Spre GARCH-prediction for sigma(t)"
Spre<- zgarch$fitted.values                     #$vector of dim N
"GARCH(p,q)-prediction, last ten values"; Spre[(N-9):N]
#Shat emp. estimator for sigma(t), moving blocks of length ka
ka<- 5; Shat<- rep(1,times=N)
for(t in ka:N) {Shat[t]<- sqrt(var(Y[(t-ka+1):t]))}
"Sigma(t)-estimation, last ten values"; Shat[(N-9):N]

#----------GARCH-residuals epsilon----------------------------
res<- Y/Spre; eps<- res[mc:N]
c("Mean epsilon"=mean(eps),"Stdev epsilon"=sqrt(var(eps)))
racf<- acf(eps,lag.max=8,type="corr",plot=F)
"Auto-correlations of GARCH-residuals epsilon"; racf$acf[1:8] #$

#--------Plot-------------------------------------------------
postscript(file="C:/CLIM/GARCHmod.ps",height=6,width=16,horiz=F)
cylim<-c(-1.5,1.5); ytext<- "Temperature [C]"
plot(Year,Y,type="l",lty=1,xlim=c(1780,2010),
                     ylim=cylim, xlab="Year",ylab=ytext,cex=1.3)
title(main=quot)
lines(Year[mc:N],Spre[mc:N],type="l",lty=1)
lines(Year[mc:N],Shat[mc:N],type="l",lty=2)
abline(h=c(-1,0,1),lty =3)

dev.off()
```

Supplement

```
garchpr<- function(y,n,a,ma,b,mb,sde){          #GARCH prediction
ys<- rep(sde,times=n)                           #vector of dim n
mc<- pmax(ma,mb)+1
for (t in mc:n){
suma<- a[1]; sumb<- 0                           #a[1] constant term
for (m in 1:(mc-1)){
suma<- suma + a[m+1]*y[t-m]^2
sumb<- sumb + b[m]*ys[t-m]^2}
ys[t]<- sqrt(suma+sumb)}
return(ys)                                      #return prediction vector ys
}

a<- rep(0,times=5); b<- rep(0,times=4)
for (m in 1:(ma+1))  {a[m]<- zgarch$coef[m]}    #$a[1] constant
if (mb > 0) {for (m in 1:mb)  {b[m]<-zgarch$coef[m+ma+1]}}   #$

"Calculation of GARCH-prediction per user function garchpr"
```

```
spre<- garchpr(Y,N,a,ma,b,mb,sde)
"GARCH(p,q)-prediction, last ten values, user function"
spre[(N-9):N]                                        #spre[ ] = Spre[ ]
```

Output from R 4.2 GARCH-results for the residual series from the ARIMA(2,2)-prediction. Hohenpeißenberg 1781–2010.

```
"Hohenpberg, Temp 1781-2010, Residuals from trend"
     Mean Y      Stdev Y      Number Years
     0.0060      0.76663      230


     ma mb mc
      3  1  4                Model: GARCH(1,3)
   Coefficient(s):
     Estimate  Std. Error  t value Pr(>|t|)
a0 4.948e-01   1.071e+01    0.046    0.963
a1 2.598e-02   1.070e-01    0.243    0.808
a2 7.034e-02   5.804e-01    0.121    0.904
a3 1.192e-14   1.495e+00 7.97e-15    1.000
b1 5.979e-02   2.031e+01    0.003    0.998

"GARCH(p,q)-prediction, last ten values"
0.7538 0.7907 0.7462 0.786 0.7737 0.7336 0.739 0.755 0.7444 0.7313
"Sigma(t)-estimation, last ten values"
0.4752 0.516 0.527 0.6676 0.6522 0.647 0.5833 0.4669 0.3883 0.7276


    Mean epsilon      Stdev epsilon
     0.00506           1.00802
  "Auto-correlations of GARCH-residuals epsilon"
  0.01264 0.00718 -0.01449  0.0278 -0.0519 -0.0205 -0.0096 -0.0567
```

4.4 Yearly Precipitation Amounts

In what follows, $Y(t)$ denotes the precipitation amount in the year t. From Y we pass to the series X by building differences, where $X(t) = Y(t) - Y(t-1), t = 2, \ldots, N$, $X(1) = 0$, see Figs. 4.4 and 4.5 (upper plots).

Table 4.5 shows that the mean yearly change X equals ≈ 0, and has an average deviation (from the mean 0) of $\approx 1.4\ldots2.0$ (dm). The auto-correlations $r(1)$ lie in the range $-0.4 \ldots -0.5$. An increase of precipitation is immediately followed by a decrease, by tendency, and vice versa.

We fit an ARMA(p,q)-model to the differenced series X. As order numbers we get $p = 3, q = 1$ and therefore the ARMA(3,1)-model

$$X(t) = \alpha_3 X(t-3) + \alpha_2 X(t-2) + \alpha_1 X(t-1) + \beta_1 e(t-1) + e(t). \quad (4.11)$$

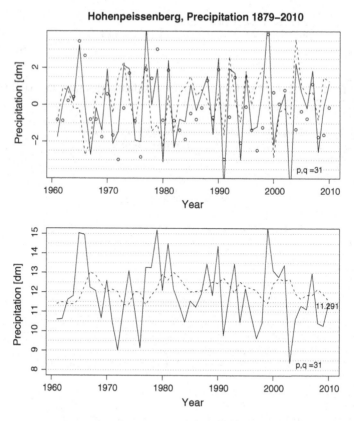

Fig. 4.4 Hohenpeißenberg, annual precipitation amounts 1879–2010. *Top* Differenced time series, together with the ARMA-prediction (*dashed line*) and with residual values (as *circles o*). *Bottom* Time series of annual precipitation amounts (dm), together with the ARIMA-prediction (*dashed line*). The last 50 years are shown

Once again, we have chosen $t_0 = [N/5]$ as starting time point for estimating the coefficients $\alpha_i = \alpha_i^{[1,t-1]}$ and $\beta_1 = \beta_1^{[1,t-1]}$, $t = t_0+1, \ldots, N$. They were used to calculate the prediction $\hat{X}(t)$ from $t_0 + 1$ onwards. Table 4.6 presents the estimated coefficients $\alpha_i = \alpha_i^{[1,N]}$ and $\beta_1 = \beta_1^{[1,N]}$ for the whole series; the coefficient β_1 is significantly different from zero (and that for all four stations)—but only one single α_i (α_2) at one single station (Po.). Further, the prognoses $\hat{Y}(t)$ for the three years 2008–2010 as well as for the year 2011 are listed, each time on the basis of the preceding years. For Karlsruhe, we have predictions for the years 2006–2009 instead of 2008–2011.

The ARIMA-prediction changes steadily from above to below the actually observed value, see Figs. 4.4 and 4.5 (lower plots). In other words: The precipitation time series oscillates heavily around a medium line (built by the predictions). Note that we had recently some relatively dry years (e.g., Bremen 2010, Hohenp. 2008 and 2009), which are overestimated by the prediction. The residuals $e(t) = Y(t) - \hat{Y}(t)$ from the predictions are shown in the upper plots of these figures.

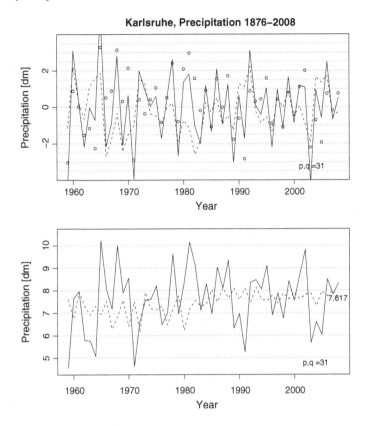

Fig. 4.5 Karlsruhe, annual precipitation amounts 1876–2008. Legend as in Fig. 4.4

Table 4.5 Differences X of precipitation amounts (dm) in consecutive years

Station	N	Mean	Standard deviation	$r(1)$	$r(2)$	$r(3)$
Bremen	121	−0.006	1.464	−0.397	−0.167	−0.037
Hohenpeißenberg	132	0.005	2.072	−0.458	−0.002	0.030
Karlsruhe	133	0.014	1.900	−0.430	−0.114	0.025
Potsdam	118	0.013	1.406	−0.462	−0.068	−0.033

Mean value, standard deviation and the first 3 auto-correlation coefficients of X are given

The auto-correlations $r_e(h)$, $h = 1, \ldots, 8$, of the residuals were calculated (but not reproduced in a Table). The bound b_1 for an individual $|r_e(h)|$ is exceeded in no case (significance level 0.05). The residual series $e(t)$ can be comprehended as a pure random series, confirming the applied ARIMA-model. We abstain here from a GARCH application to the residual series.

As in Sect. 4.2, we compare the ARIMA-method with the left-sided moving averages, see Table 4.7. The latter needs a depth of 4 (Ho: 2; Ka: 6) to beat the former method. This means that here—in the case of annual precipitation—the method of left-sided moving averages is on a par with the ARIMA approach.

Table 4.6 ARIMA-method for the annual precipitation amounts (dm)

sd	Order p, q	ARMA-coefficients α_i	β_1	Root MSQ	ARIMA-prediction 2008–2010	2011
B	3, 1	−0.096, −0.238, −0.117	−0.873	1.178	7.51, 7.12, 7.15	7.30
1.408	obs:				7.00, 6.45, 5.60	6.22
H	3, 1	0.115, 0.077, 0.012	−0.930	1.900	12.17, 11.91, 11.57	11.66
2.203	obs:				10.42, 10.28, 11.43	12.47
P	3, 1	−0.128, −0.094, −0.040	−0.995	1.126	5.87, 5.95, 6.01	6.09
1.409	obs:				5.75, 5.98, 6.58	6.20
					2006–2008	2009
K	3, 1	0.046, −0.163, −0.062,	−0.943	1.530	7.76, 8.00, 7.58	7.61
1.950	obs:				8.51, 7.83, 8.33	–

Coefficients, goodness-of-fit, predictions for the years 2008–2011 (Karlsruhe: 2006–2009), with actually observed values beneath. Further: sd = sd(X) denotes the standard deviation of the $N_0 = N - t_0$ values of the differenced series $X(t), t = t_0 + 1, \ldots, N$. B Bremen, H Hohenpeißenberg, K Karlsruhe, P Potsdam

Table 4.7 Left-sided moving averages for annual precipitation amounts

Depth k	RootMSQ Bremen	Hohenpeißenberg	Karlsruhe	Potsdam
3	1.260	1.877	1.662	1.186
4	1.156	1.895	1.567	1.116
5	1.120	1.883	1.533	1.106
6	1.131	1.884	1.527	1.077
7	1.129	1.907	1.516	1.053
8	1.129	1.901	1.478	1.058
ARIMA (p, q)	(3, 1) 1.178	(3, 1) 1.900	(3, 1) 1.530	(3, 1) 1.126

Depth k of averaging and the resulting goodness-of-fit RootMSQ are listed. The latter is given for the ARIMA-method, too (see Table 4.6)

We derive the standardized RootMSQ measures (4.7) for annual precipitation from Table 4.6,

Bremen 0.836,   Hohenpeißenberg 0.862,   Karlsruhe 0.784,   Potsdam 0.800.

These rsq-values do not differ much from station to station, nor do they differ much from the rsq-values in Sect. 4.2 for temperature.

GARCH-Modeling the Annual Precipitation

The poor significance of the AR-coefficients α_i in Table 4.6 corresponds with the result of Table 3.7. It leads us once again to the question, whether the yearly precipitation series $Y(t)$ is (close to) a pure random process. Let us suppose now, that $Y(t)$ forms a series of uncorrelated variables, having a conditional variance $\sigma^2(t)$ – given the information up to time $t - 1$. We build the centered process $X(t) = Y(t) - \bar{Y}$

Table 4.8 GARCH-modeling of the annual precipitation amounts (dm)

GARCH-coefficients		Mean	Auto-correlations r_e	Max	Bound	
$\alpha_1, \alpha_2, \alpha_3$	β_1	$\hat{\sigma}(t)$	$r_e(1), r_e(2), r_e(3)$	$|r_e(h)|$	b_1	b_8
B 0.000, 0.021, 0.127	0.098	1.069	0.059, −0.097, −0.016	0.203	0.180	0.252
H 0.051, 0.000, 0.044	0.081	1.702	0.270, 0.213, 0.156	0.270	0.173	0.242
K 0.041, 0.000, 0.070	0.051	1.325	0.002, −0.130, −0.004	0.130	0.172	0.240
P 0.094, 0.000, 0.043	0.053	0.956	−0.066, −0.150, −0.095	0.150	0.183	0.255

Coefficients α_i and β_1 are presented (the constant term α_0 is 0.871, 2.38, 1.47, 0.742 for B,H,K,P, resp.). Further, the mean of the predicted standard deviations $\hat{\sigma}(t)$, the first 3 auto-correlations of the residual series $e = X/\hat{\sigma}$, with max $|r_e(h)|$ out of the 8 values for $h = 1, \ldots, 8$, and the individual and simultaneous statistical bounds b_1 and b_8 are given. *B* Bremen, *H* Hohenpeißenberg, *K* Karlsruhe, *P* Potsdam

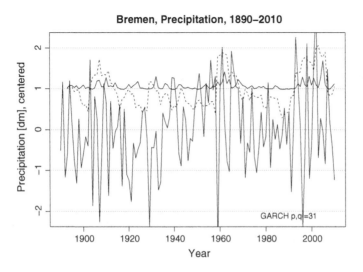

Fig. 4.6 Bremen, annual precipitation amounts 1890–2010, centered (*zigzag line*). GARCH-prediction for $\sigma(t)$ (*inner solid line*) and standard deviation of left sided moving (5-years) blocks (*dashed line*)

and write down a GARCH-model of order (3,1), that is cf. Sect. 4.3

$$X(t) = \sigma(t) \cdot \epsilon(t), \qquad t = 1, 2, \ldots$$
$$\sigma^2(t) = \alpha_3 X^2(t-3) + \alpha_2 X^2(t-2) + \alpha_1 X^2(t-1) + \alpha_0 + \beta_1 \sigma^2(t-1). \quad (4.12)$$

Other order numbers, like (2,2) or (1,3) instead of (3,1), lead to the same results. Table 4.8 brings the estimated GARCH-coefficients. By means of these estimations one calculates the GARCH-predictions $\hat{\sigma}(t)$. Note that the mean value of these predictions—see Table 4.8—is close to the standard deviation of the process $Y(t)$ ($s_Y = 1.067, 1.719, 1.353, 0.960$ for B,H,K,P, resp., acc. to Table 1.3) and close to the square root of

$$\alpha_0/(1 - \alpha_1 - \alpha_2 - \alpha_3 - \beta_1),$$

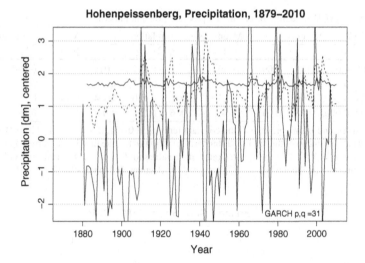

Fig. 4.7 Hohenpeißenberg, annual precipitation amounts 1879–2010, centered. Legend as in Fig. 4.6

the latter fact follows from the theory of GARCH-models. For the GARCH-residuals (see the first equation in (4.12))

$$\hat{\epsilon}(t) = e(t) = \frac{X(t)}{\hat{\sigma}(t)},$$

we obtain (mean, variance) $\approx (0,1)$, as the model demands. Further, we compute for $e(t)$ the auto-correlation function $r_e(h)$, $h = 1, \ldots, 8$. Except for Hohenpeißenberg, the $|r_e(h)|$-values are small and not significantly different from zero. Hence, for Bremen, Karlsruhe, and Potsdam, the GARCH-residuals $e(t)$ can be assumed to form a series of uncorrelated variables. So the GARCH-model seems to fit well in these cases, and this supports the assumption of uncorrelated $X(t)$ (and hence $Y(t)$), which is part of the GARCH definition.

For the precipitation series, we find out once more a proximity to the pure random series (exception: Hohenpeißenberg, see Fig. 4.7).

The coefficients α_i and β_1 in Table 4.8 are small and not significantly different from zero (all t-values smaller than 0.9). This means that we can assume a nearly constant (conditional) variance $\sigma^2(t) \approx Var(X)$. That corresponds with the GARCH-predictions $\hat{\sigma}(t)$ in Figs. 4.6 and 4.7, varying little around the horizontal line, built by the standard deviation $s_X = s_Y$.

Chapter 5
Model and Prediction: Monthly Data

For the investigation of monthly climate data, we first estimate a trend by the ARIMA- or by the moving average-method of Chap. 4. Then we remove the trend and apply the ARMA- or moving average-method once again, now to the detrended series. In Sect. 8.3 we will present a sin-/cos-approach to monthly data.

5.1 Trend+ARMA Method for Monthly Temperature Means

In order to model the monthly temperature means $Y(t)$, we start with the decomposition

$$Y(t) = m(t) + X(t), \quad t = 1, 2, \ldots, \tag{5.1}$$

where t counts the successive months, $m(t)$ denotes the long-term trend, and where $X(t)$ is the remainder series. We estimate the trend by the ARIMA-method of Sect. 4.1: The variable $m(t)$ is the ARIMA-prediction of the yearly temperature mean (see Table 4.2 and Figs. 4.1, 4.2); $m(t)$ will be called ARIMA-trend, and is the same for all 12 months t of the same year. The *detrended* series

$$X(t) = Y(t) - m(t), \quad t = 1, 2, \ldots,$$

is shown in the upper plots of Figs. 5.1 and 5.2. We fit an ARMA(p, q)-model to the series $X(t)$, with $p = 3, q = 2$ for Hohenpeißenberg, and $p = 2, q = 3$ for Bremen, Karlsruhe and Potsdam (these p, q-values finally turned out to be sufficiently large). In the latter case, we are faced with the model

$$X(t) = \alpha_2 X(t-2) + \alpha_1 X(t-1) + \beta_3 e(t-3) + \beta_2 e(t-2) + \beta_1 e(t-1) + e(t).$$

To fit the model, we estimate coefficients $\alpha_i = \alpha_i^{[1,t-t]}$, $\beta_j = \beta_j^{[1,t-1]}$ (for each month t anew) and calculate the prediction $\hat{X}(t)$, $t = t_0 + 1, \ldots, M$, by analogy with Sect. 4.1; $M = N * 12$ is the number of months, N the number of years. Hereby,

H. Pruscha, *Statistical Analysis of Climate Series*,
DOI: 10.1007/978-3-642-32084-2_5, © Springer-Verlag Berlin Heidelberg 2013

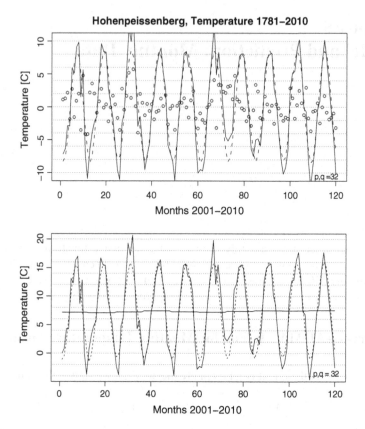

Fig. 5.1 Hohenpeißenberg, monthly temperature means 1781–2010. *Top* Detrended time series, together with the ARMA-prediction (*dashed line*) and with the residual values (as *circles o*). *Bottom* Monthly temperature means (°C), together with the ARIMA-trend (*inner solid line*) and the trend+ARMA-prediction (*dashed line*). The last 10 years are shown

we choose $t_0 = [N/5] * 12$ as starting month for the prediction. In Table 5.1 one can find the estimated coefficients $\alpha_i = \alpha_i^{[1,M]}$ and $\beta_j = \beta_j^{[1,M]}$; (nearly) all of them are significantly different from zero.

If we consider the AR(2) part of the ARMA-model separately and apply Eq. (B.10) of the Appendix in order to find the period T, where the spectral density is maximal, we get successively

$$\cos(\omega) = 0.866, \quad \omega = 0.5236, \quad T = 2 \cdot \pi/\omega = 12.00 \text{ (months)}$$

for Bremen, Karlsruhe and Potsdam (T = 11.54 for Hohenp.). The yearly periodicity of temperature is correctly reproduced by our ARMA-model.

The ARMA-predictions $\hat{X}(t)$ for $X(t)$ are plotted in the upper parts of Figs. 5.1, 5.2, too. By means of $\hat{X}(t)$ we gain back the original (trend-affected) series, more

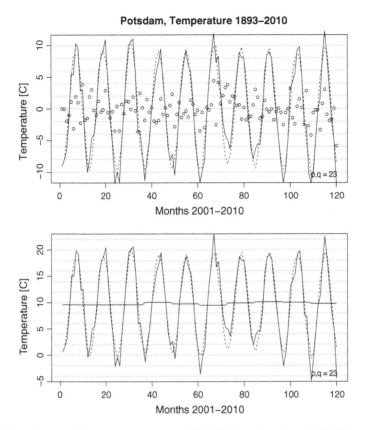

Fig. 5.2 Potsdam, monthly temperature means 1893–2010. Same legend as in Fig. 5.1

precisely: the ARIMA-trend+ARMA-prediction $\hat{Y}(t)$ for $Y(t)$. We put

$$\hat{Y}(t) \;=\; m(t) + \hat{X}(t), \quad t = 1, 2, \ldots, \tag{5.2}$$

compare the lower plots of Figs. 5.1, 5.2, where the predictions $\hat{Y}(t)$ are portrayed, together with the actual observations $Y(t)$. Table 5.1 presents the goodness-of-fit values *RootMSQ* according to Eq. (4.6)—replacing N by $M = N * 12$—and the predictions for Oct. 2010 to Jan. 2011 (Karlsruhe: Oct. 2008 to Jan. 2009). With only 4+5 parameters these ARIMA-trend+ARMA-predictions $\hat{Y}(t)$ run close to the actual observed values $Y(t)$. They cannot, however, follow extremely warm summers or cold winters. To give examples, we point to the "record summer" 2003, mentioned in Sect. 2.5 above (in Fig. 5.1 around the month no. 31), or to the relatively cold months Jan. 2009, Jan. 2010, Dec. 2010. For the latter, compare the predictions 2.49, 0.63, 1.46 (°C) with the actual observed values -3.3, -2.5, -4.4 (°C) in Bremen, Hohenpeißenberg, and Potsdam, respectively (Table 5.1).

Table 5.1 The ARIMA-trend+ARMA-method for monthly temperature means (°C)

sd	Order p, q	ARMA-coefficients α_i	β_j	Root MSQ	Prediction Oct–Dec 2010	Jan 2011
B	2, 3	1.732, −1.000	−1.40, 0.50, 0.24	1.906	10.22, 6.01, 2.49	−1.17
6.242	obs:				9.4, 4.9, −3.3	2.3
H	3, 2	1.77, −1.07, 0.04	−1.726, 0.994	2.141	8.37, 4.04, 0.63	−1.33
6.497	obs:				6.8, 3.1, −2.5	−1.0
P	2, 3	1.732, −1.00	−1.48, 0.58, 0.23	2.013	9.57, 4.63, 1.46	−1.91
7.103	obs:				7.9, 4.7, −4.4	1.1
					Oct–Dec 2008	2009
K	2, 3	1.732, −1.00	−1.51, 0.63, 0.18	1.909	11.11, 6.81, 3.70	2.13
6.808	obs:				10.9, 7.3, 2.4	−1.3

Coefficients, goodness-of-fit, prediction for Oct.–Dec. 2010, Jan. 2011 (Karlsruhe Oct.–Dec. 2008, Jan. 2009), with the actually observed value beneath. Further: sd denotes the standard deviation of the $N * 12 - t_0$ values of the detrended series. B Bremen, H Hohenpeißenberg, K Karlsruhe, P Potsdam

As in Eq. (4.7), we build the standardized *RootMSQ* measure

$$\text{rsq} = RootMSQ/\text{sd}(X),$$

with the standard deviation sd(X) of the $N * 12 - t_0$ monthly values of the detrended series X. We obtain rsq =

$$0.305 \text{ (Bremen)}, \quad 0.330 \text{ (Hohenp.)}, \quad 0.280 \text{ (Karlsr.)}, \quad 0.283 \text{ (Potsd.)}, \qquad (5.3)$$

which is about the same level for the four stations.

R 5.1   Monthly temperature data. Trend removal, ARMA-modeling for the detrended series, prediction for the detrended series and for the (original) series with trend, residual analysis. The trend `Ytr` is the yearly trend according to 4.2, stored in the file `PoT40Pre.txt`. For the sake of simplicity, the method of standard regression analysis, mentioned in 4.2, Remark A above, is adopted to monthly data and applied here: The coefficients are estimated only once (from the whole series) and then used for predicting the detrended series and the (original) series with trend, by means of `Ydfit[t]` and `YIarma[t]`, resp.

```
library(TSA)                              #see Cran-sofware-packages
attach(potsdTp)
#-------Data preparation, trend removal----------------------
"Monthly temperature means Potsdam 1893-2010"
mon12<- data.frame(potsdTp[,3:14])/10          #selecting jan-dec
NYear<- length(Tyear);   M<- NYear*12
c("Number Years"=NYear,"Number Months M"=M)
detach(potsdTp)
```

```
#Read the yearly trend Ytr, twelvefold as Ytre
potsdPr<- read.table("C:/CLIM/PoT40tPr.txt",header=T)
attach(potsdPr)                         #contains variable Ytr
Ytre<- 1:M                              #Ytre vector of dim M
for(m in (1:M)){j<- trunc((m-1)/12)+1;Ytre[m]<- Ytr[j]}
#Instead of the last two lines:
#twelve<- rep(12,times=NYear); Ytre<- rep(Ytr,twelve)
YtreN1<- 9.2743       #Forecast New Year from R 4.1 (Table 4.2)

Yobs<- as.matrix(t(mon12))  #t=transpose, as 12 x NYear matrix
dim(Yobs)<- c(M,1)                      #as M-dim vector
Yde<- Yobs - Ytre                       #detrending
ma<- 2; mb<- 3 ;  mc<- pmax(ma,mb)      #ARMA order <= 6
tst<- trunc(NYear/5)*12; ts1<-tst+1     #Start for prediction

#------ARMA-model for detrended monthly data----------------
#---a) Estimation---------------------------
Ydarma<- arma(Yde,order=c(ma,mb))
Ydcoef<- Ydarma$coef;    Yds2<- Ydarma$css
Ydfit<- Ydarma$fitted.values;    Ydres<- Ydarma$residuals
c("Start at"=tst,"ArOrder"=ma,"MAOrder"=mb,"Cond.SSQ"=Yds2)
summary(Ydarma)

a<- rep(0,times=6); b<- rep(0,times=6)
if (ma > 0) for (m in 1:ma){a[m]<- Ydcoef[m]}
if (mb > 0) for (m in 1:mb){b[m]<- Ydcoef[ma+m]}
theta<- Ydcoef[ma+mb+1]                         #intercept

#---b) Prediction-----------------------------
"ARMA-prediction for the detrended series, last 12 months"
                                  Ydfit[(M-11):M]
YdarmaNJ<- theta
for (m in 1:mc) {YdarmaNJ<-
        YdarmaNJ+a[m]*Yde[M+1-m]+b[m]*Ydres[M+1-m]}
c("Forecast Jan_NewYear, detrended series"=YdarmaNJ)
# (Original) series with trend
YIarma<- Ytre; YIarma[ts1:M]<- Ytre[ts1:M]+Ydfit[ts1:M]
Yres<- Yobs - YIarma
"ARMA-prediction for the series with trend,last 12 months"
                                  YIarma[(M-11):M]
"Forecast Jan_NewYear, series with trend"
c(YtreN1,YdarmaNJ,YdarmaNJ+YtreN1)

#---c) Residual analysis--------------------
Yrer<- Yres[ts1:M]; msq<- mean(Yrer*Yrer)
c("Mean Yres"=mean(Yrer),"Std Yres"=sqrt(var(Yrer)),
  "MSQ"=msq,"RootMSQ"=sqrt(msq))
racf<-acf(Yrer,lag.max=8,type="corr",plot=F)$acf   #$no Plot
"Autocorrelation of residual series"; racf[1:8]
```

Output from R 5.1 ARMA-output for the detrended series, prediction for the detrended series and for the (original) time series with trend, residual analysis with *RootMSQ*-value and auto-correlation function. The standard regression approach was chosen. With the exception of $r_e(1) = 0.031$ instead of 0.001, see Table 5.3, the results here are similar to those from the forecast regression approach. Notice that Cond.SSQ/(M-6) = 3.84 ≈ MSQ.

```
"Monthly temperature means Potsdam 1893-2010"
  Number Years       Number Months M
          118              1416
  Start at    ArOrder    MAOrder    Cond.SSQ
     276         2          3          5420


Model: ARMA(2,3)
Coefficient(s):
             Estimate   Std. Error   t value  Pr(>|t|)
ar1          1.732206     0.000211   8209.70   <2e-16 ***
ar2         -1.000223     0.000210  -4759.75   <2e-16 ***
ma1         -1.476013     0.023814    -61.98   <2e-16 ***
ma2          0.578910     0.040879     14.16   <2e-16 ***
ma3          0.230042     0.024017      9.58   <2e-16 ***
intercept    0.004332     0.017266      0.25      0.8

sigma^2 estimated as 3.84, Cond.Sum-of-Squares = 5420, AIC = 5935


"ARMA-prediction for the detrended series, last 12 months"
-10.154 -9.419 -5.022 -0.077 4.756 7.539 9.771 9.181 4.762 -0.287
-5.171 -8.349
Forecast Jan_NewYear, detrended series              -11.185


 "ARMA-prediction for the series with trend, last 12 months"
-0.365 0.371 4.768 9.712 14.545 17.328 19.560 18.970 14.551 9.502
 4.619 1.440
 "Forecast Jan_NewYear, series with trend"  9.274-11.185 = -1.911
  Mean Yres   Std Yres    MSQ      RootMSQ
  -0.0291     2.0009    4.0010     2.0003


 "Autocorrelation of residual series"
0.03114 0.12581 0.04704 0.04800 0.03702 0.02462 0.08007 0.05606
```

5.2 Comparisons with Moving Averages and Lag-12 Differences

On the basis of approach (5.1), we can alternatively choose the method of left-sided moving averages, applied twofold, for estimating the trend $m(t)$ as well as for predicting the detrended series $X(t)$. We estimate the trend $m(t)$ by building the average of the preceding observations

Table 5.2 The MA-trend+MA-method for monthly temperature means

Depth (years)	RootMSQ			
k	Bremen	Hohenp.	Karlsruhe	Potsdam
5	2.143	2.343	2.091	2.294
10	2.013	2.217	1.963	2.141
15	1.973	2.178	1.929	2.103
20	1.950	2.154	1.906	2.083
25	1.931	2.141	1.899	2.061
ARIMA-trend+ARMA	1.906	2.141	1.909	2.013
ARIMA(lag12)	2.181	2.543	2.301	2.325

Depth k of the left-sided moving average and the resulting goodness-of-fit values *RootMSQ*. The latter is listed for the ARIMA-trend+ARMA-method (cf. Table 5.1) and for the ARIMA(lag12)-method, too

$$Y(t-1), Y(t-2), \ldots, Y(t-k*12)$$

(MA-trend). The averaging comprises $k*12$ months, k indicates the number of the employed years ("depth"). As prediction $\hat{X}(t)$ for the variable $X(t)$, we take the average of the preceding *detrended* observations

$$Y(t-12) - m(t-12), Y(t-24) - m(t-24) \ldots, Y(t-k*12) - m(t-k*12).$$

Here, the average is taken over the (detrended) climate values of the same calendar month in k preceding years. Again, we gain the prediction $\hat{Y}(t)$ for $Y(t)$ according to Eq. (5.2). Then, by Eq. (4.6), with N replaced by $M = N*12$, we compute the goodness of this prediction method (called MA-trend+MA-method). The starting month for prediction is once again $t_0 = [N/5]*12$. Table 5.2 shows that for no depth smaller than $k = 19$ (years) the *RootMSQ* values of the ARIMA-trend+ARMA-method are attained. Recall that the latter prediction method needs only $4+5=9$ lagged variables, thus seeming to be superior to the MA-trend+MA-method.

Another alternative procedure resembles the ARIMA-method of Sect. 4.1. Instead of using differences $Y(t) - Y(t-1)$ of two consecutive variables (lag-1 differences), however, we form lag-12 differences that are differences

$$X(t) = Y(t) - Y(t-12), \quad t = 13, 14, \ldots,$$

of two observations being separated by twelve months. We fit an AR(12)-model to this differenced process $X(t)$, and determine the goodness-of-fit by Eq. (4.5) or—equivalently—Eq. (4.6), with N replaced by M. We will use the short-hand notation ARIMA(lag12). Table 5.2 shows that this procedure is inferior to the method ARIMA-trend+ARMA and to the method MA-trend+MA as well.

Table 5.3 Auto-correlation function $r_e(h)$, up to time lag $h = 8$ (months), of the ARIMA-trend+ARMA-residuals; together with individual and simultaneous bounds b_1 and b_8, resp. [level 0.01]

	$r_e(1)$	$r_e(2)$	$r_e(3)$	$r_e(4)$	$r_e(5)$	$r_e(6)$	$r_e(7)$	$r_e(8)$	b_1	b_8
Br	0.005	0.083	0.039	0.04	0.06	0.06	0.10	0.07	0.075	0.095
Ho	0.029	0.039	−0.014	−0.00	0.00	0.03	0.01	−0.01	0.055	0.069
Ka	−0.061	0.067	−0.022	0.02	0.06	0.06	0.05	0.02	0.057	0.072
Po	0.001	0.135	0.045	0.06	0.04	0.03	0.08	0.06	0.076	0.096

Monthly temperature means at the stations *Br* Bremen, *Ho* Hohenpeißenberg, *Ka* Karlsruhe, *Po* Potsdam

Table 5.4 The ARIMA-trend+ARMA-method for the monthly precipitation amounts (cm)

sd	Order p, q	ARMA-coefficients α_i	β_j	Root MSQ	Prediction Oct–Dec 2010	Jan 2011
B	2,2	0.624, −0.245	−0.496, 0.229	3.104	6.23, 5.63, 5.64	4.56
3.106	obs:				3.71, 5.62, 3.26	3.48
H	2, 2	1.732, −0.999	−1.707, 0.971	4.710	9.30, 6.46, 4.71	4.15
5.955	obs:				6.24, 4.54, 6.96	5.32
P	4, 0	0.12, 0.05, −0.04, −0.00		2.945	6.12, 4.66, 4.91	5.59
2.946	obs:				1.84, 9.28, 7.28	3.90
					Oct–Dec 2008	2009
K	3, 1	−0.200, 0.07, 0.02	0.295	3.621	6.36, 6.77, 6.53	6.45
3.632	obs:				10.70, 6.04, 6.28	–

Coefficients, goodness-of-fit, predictions for Oct.–Dec. 2010, Jan. 2011 (Karlsruhe Oct.–Dec. 2008, Jan. 2009), with actually observed values beneath. Further: sd denotes the standard deviation of the $N * 12 - t_0$ values of the detrended series. *B* Bremen, *H* Hohenpeißenberg, *K* Karlsruhe, *P* Potsdam

5.3 Residual Analysis: Auto-Correlation

Denoting once again by $M = N * 12$ the total number of months and by $\hat{Y}(t)$ the ARIMA-trend+ARMA–prediction for $Y(t), t = t_0 + 1, \ldots, M$, we compute by

$$e(t) = Y(t) - \hat{Y}(t), \quad t = t_0 + 1, \ldots, M,$$

the residuals from the prediction; compare Sect. 5.1 and the upper plots in Figs. 5.1 and 5.2. Which structure has this residual time series $e(t), t = t_0 + 1, \ldots, M$? Its auto-correlation function $r_e(h), h = 1, \ldots, 8$, consists of values more or less near zero, cf. Table 5.3, with the exceptions of $r_e(7)$ (Bremen) and $r_e(2)$ (Potsdam). It is particularly the auto-correlation $r_e(1)$ of first order (that is the correlation between $e(t), e(t + 1)$ in two immediately succeeding months), which turns out to be satisfactorily small. The simultaneous bound $b_8 = u_{1-0.005/8}/\sqrt{M_0}$, see also Sect. 3.3, is exceeded (little) in the two cases mentioned above. Due to the large values of $M_0 = M - t_0$, we choose the significance level 0.01 instead of 0.05. The prediction method ARIMA-trend+ARMA leaves behind residuals, which are little correlated and

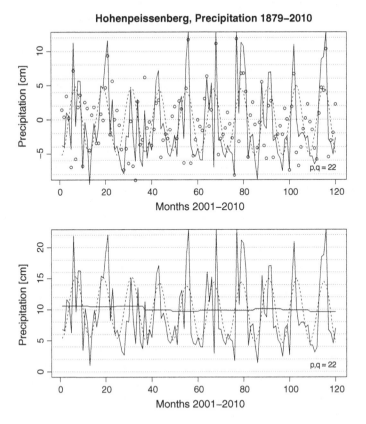

Fig. 5.3 Hohenpeißenberg, monthly precipitation amounts, 1879–2010. *Top* Detrended time series, together with the ARMA-prediction (*dashed line*) and with the residual values (as *circles o*). *Bottom* Monthly precipitation amounts (cm), together with the ARIMA-trend (*inner solid line*) and the trend+ARMA-prediction (*dashed line*). The last 10 years are shown

thus do fulfill the demand on residual variables $e(t)$. This statement is not true with respect to the method MA-trend+MA of moving averages (Sect. 5.2). Here, for all four stations, the auto-correlations $r_e(1)$ are too large, namely for a depth of $k = 15$,

$$r_e(1) = 0.284, 0.112, 0.144 \text{ and } 0.262 \text{ for Br, Ho, Ka and Po, resp.}.$$

5.4 Monthly Precipitation Amounts

Table 5.4 is dedicated to monthly precipitation amounts and constructed by analogy to Table 5.1 for monthly temperature means. Some predictions are far away from the actual observed values. Figures 5.3 and 5.4 reveal the reason: Our forecast procedures cannot cope with the large (random) oscillations of the monthly precipitation

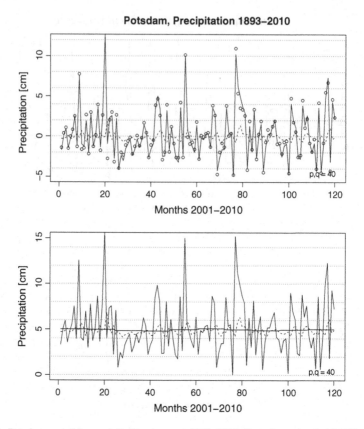

Fig. 5.4 Potsdam, monthly precipitation amounts, 1893–2010. Same legend as in Fig. 5.3

series. This is especially true for Potsdam (and Bremen, Karlsruhe, no plots), and is somewhat weaker the case for Hohenpeißenberg. This statement is confirmed when forming the standardized *RootMSQ* measure rsq. With the standard deviation sd(X) of the $N * 12 - t_0$ values of the detrended series X, we have rsq = $RootMSQ$/sd(X) =

$$0.999 \text{ (Bremen)}, \quad 0.791 \text{ (Hohenp.)}, \quad 0.996 \text{ (Karlsr.)}, \quad 0.999 \text{ (Potsd.)}. \qquad (5.4)$$

As expected, the rsq value of Hohenpeißenberg is smaller than that of the other three stations; their value near 1 indicates a nearly total indetermination. But all four values (5.4) are much larger than the corresponding values (5.3) for monthly *temperature*, showing once more (after Sect. 3.1) that the process of monthly precipitation is much more irregular than that of monthly temperature (see Sect. 8.4 for a further discussion).

Chapter 6
Analysis of Daily Data

We start with basic informations on the daily temperature and precipitation data from five German stations, collected over the years 2004–2010; see Appendix A.3. Besides Bremen, Hohenpeißenberg and Potsdam, see Table 1.1, we include in our analysis: The westernmost German city Aachen (202 m, 50°47', 06°05') in the middle Rhineland and Würzburg (268 m, 49°46', 09°57') upon the river Main.

The 29th Feb 2004 and 2008 were canceled (to produce plots like Figs. 6.1 or 6.7). Thus, we have selected 7*365 = 2555 days. We had to drop the station Karlsruhe covering the years until 2008 only.

First, we are interested in the spatial aspect of the data. That is the question, how the observations at the single stations are cross-correlated. With respect to temperature, we employ the cross-correlation function from time series analysis. When dealing with precipitation, methods from categorical data analysis seem to be more appropriate. The reason is, that we have many days without any precipitation (about half the days, see Table 6.1). Here, logistic regression and contingency table analysis are applied.

Days with heavy rainfall are treated as *rare* events and are investigated with methods of event-time analysis. Here, we use intensity functions and the model of an inhomogeneous Poisson process.

6.1 Series of Daily Climate Records

Table 6.1 shows that the temperature series have a high auto-correlation $r(1)$, which decreases slightly (by 0.1), when seasonally adjusted. The precipitation series have a small $r(1)$ value, staying nearly the same after adjustment.

As it is the case with yearly precipitation (see Tables 3.1, 3.2, 3.3), Hohenpeißenberg's daily precipitation series has the largest auto-correlation coefficients. Thus it may contain more inner correlation structure than the others.

H. Pruscha, *Statistical Analysis of Climate Series*,
DOI: 10.1007/978-3-642-32084-2_6, © Springer-Verlag Berlin Heidelberg 2013

Table 6.1 Descriptive measures of the daily temperature (°C) and precipitation (mm) data for five stations and seven years 2004–2010; mean m, standard deviation sd, auto-correlations $r(1)$ and $r_e(1)$ of order 1, without and with seasonal adjustment, respectively

Station	Temperature				Precipitation				
	m	sd	$r(1)$	$r_e(1)$	W %	m	sd	$r(1)$	$r_e(1)$
Aachen	10.58	6.91	0.949	0.841	47.0	2.26	4.60	0.149	0.148
Bremen	9.82	6.93	0.951	0.837	48.6	1.84	3.87	0.166	0.161
Hohenpeißenberg	7.45	7.85	0.929	0.813	48.3	3.06	6.54	0.263	0.240
Potsdam	9.69	7.86	0.959	0.850	51.8	1.70	3.79	0.164	0.160
Würzburg	9.99	7.73	0.960	0.855	52.1	1.69	4.02	0.181	0.175

W % stands for the percentage of days without precipitation

For the sake of clearness, the plots over the calendar days are restricted to the four years 2004–2007 (thus presenting four points for each day). The course of temperature over the year possesses a strong seasonal component, see upper parts in Figs. 6.1, 6.2, 6.3. This is different from precipitation (lower plots), where we observe only a weak summer effect (which is stronger at Hohenpeißenberg); compare also Fig. 2.6 for monthly temperature and precipitation.

As a consequence, we will analyze the daily temperature data only in the seasonally adjusted form (in the next section). This is done by removing a polynomial(4) spanned over the 365 calendar days (see upper plots of Figs. 6.1, 6.2, 6.3). The precipitation data, however, will be let unaltered (in Sects. 6.3 and 6.4).

6.2 Temperature: Cross-Correlation Between Stations

Let us assume that we have the bivariate sample

$$(X_1, Y_1), (X_2, Y_2), \ldots, (X_n, Y_n) \qquad (6.1)$$

of two time series X_t and Y_t. We gain an estimator $c_{xy}(h)$ for the cross-covariance $\gamma_{xy}(h)$ (see Appendix B.1) according to the following equations. For positive and for negative time lags h we put

$$c_{xy}(h) = \begin{cases} \frac{1}{n} \sum_{t=1}^{n-h} (X_t - \bar{X})(Y_{t+h} - \bar{Y}) & \text{for } h = 0, 1, 2, \ldots, \\ \frac{1}{n} \sum_{t=1+|h|}^{n} (X_t - \bar{X})(Y_{t-|h|} - \bar{Y}) & \text{for } h = -1, -2, \ldots. \end{cases}$$

Hereby, $|h|$ should not exceed $[n/4]$, acc. to Box & Jenkins (1976), and

$$\bar{X} = \frac{1}{n} \sum_{t=1}^{n} X_t \quad \text{and} \quad \bar{Y} = \frac{1}{n} \sum_{t=1}^{n} Y_t$$

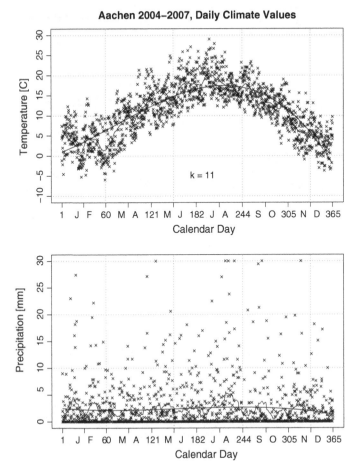

Fig. 6.1 Daily Temperature (*top*) and Precipitation (*bottom*) in Aachen 2004–2007, plotted over the 365 Calendar Days. A fitted polynomial of order 4 (*smooth line*) and a 11-days moving average (*oscillating line*) are drawn. The precipitation amount is truncated at 30 mm

denote the mean values of the x- and y-sample, resp. From the coefficients $c_{xy}(h)$ we get the empirical cross-correlation function or *cross-correlogram* $r_{xy}(h)$, $h = 0$, $\pm 1, \pm 2, \ldots$, by

$$r_{xy}(h) = \frac{c_{xy}(h)}{s_x \cdot s_y}, \qquad s_x = \sqrt{c_{xx}(0)}, \; s_y = \sqrt{c_{yy}(0)};$$

s_x and s_y are the standard deviations of the x- and y-sample. We have

$$c_{xy}(-h) = c_{yx}(h), \quad r_{xy}(-h) = r_{yx}(h), \quad |r_{xy}(h)| \le 1;$$

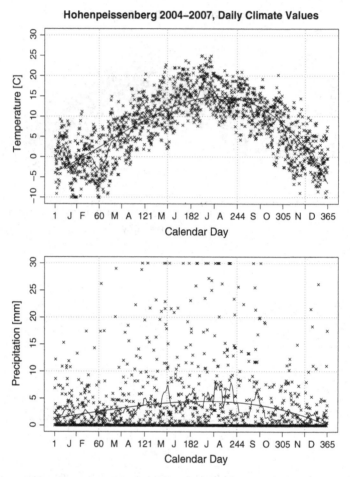

Fig. 6.2 Daily Temperature (*top*) and Precipitation (*bottom*) at Hohenpeißenberg 2004–2007. Same legend as in Fig. 6.1

$r_{xy}(0)$ is the usual correlation coefficient of the bivariate sample (6.1), and $r_{xx}(h) = r_x(h)$ is the auto-correlation of the x-sample with time lag h, as used in Sect. 3.3.

Table 6.2 presents cross-correlograms $r_{xy}(h)$ for daily temperature (seasonally adjusted), with $x =$ Aachen (and then with $x =$ Potsdam), and with $y =$ Aachen,..., Würzburg, for positive time lags $h = 0, \ldots, 8$ (days).

Daily temperatures are positively correlated between all stations, but with decreasing values for increasing time lags. The r_{xy}-values for Aachen $\to$ Potsdam and Potsdam $\to$ Aachen are presented in the first plot of Fig. 6.4. The Aachen $\to$ Potsdam values lie above those of Potsdam $\to$ Aachen, more or less clearly up to a time lag of 4 or 5 days. Note that the first curve represents the west $\to$ east,

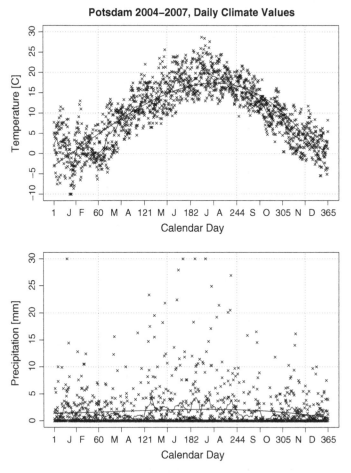

Fig. 6.3 Daily Temperature (*top*) and Precipitation (*bottom*) in Potsdam 2004–2007. Same legend
as in Fig. 6.1

the second the reversed direction. The same phenomenon can be observed for Bremen → Potsdam (and reversed) as well as for Aachen → Würzburg (and reversed).
So we can state, that the prevailing wind direction is reflected.

The cross-correlations between Bremen and Hohenpeißenberg are low and not
(strongly) affected by the choice of direction (fourth plot in Fig. 6.4). The cross-
correlations between a station and Hohenpeißenberg are smaller—from time lag 2
(Würzburg: 3) onwards—than between that station and any other station, see e.g.,
Table 6.2, demonstrating the somewhat special position of this mountain station.

Table 6.2 Cross-correlation coefficients $r_{xy}(h)$, $h = 0, \ldots, 8$ (days) for daily temperature (seasonally adjusted), in the years 2004–2010

| | $x =$ Aachen $\rightarrow$ | | | | | |
| | $y =$ | | | | | |
Lag	Aachen	Bremen	Hohenp.	Potsdam	Würzburg	Lag
0	1.000	0.885	0.816	0.815	0.857	0
1	0.841	0.809	0.773	0.813	0.839	1
2	0.644	0.642	0.603	0.680	0.691	2
3	0.513	0.518	0.462	0.556	0.557	3
4	0.424	0.445	0.368	0.472	0.464	4
5	0.362	0.390	0.298	0.412	0.401	5
6	0.315	0.348	0.253	0.360	0.347	6
7	0.275	0.311	0.222	0.323	0.307	7
8	0.240	0.279	0.199	0.291	0.273	8
	$x =$ Potsdam $\rightarrow$					
	$y =$					
Lag	Aachen	Bremen	Hohenp.	Potsdam	Würzburg	Lag
0	0.815	0.917	0.738	1.000	0.866	0
1	0.677	0.756	0.638	0.850	0.778	1
2	0.555	0.608	0.514	0.680	0.638	2
3	0.475	0.519	0.423	0.571	0.533	3
4	0.420	0.462	0.363	0.499	0.466	4
5	0.374	0.417	0.320	0.446	0.419	5
6	0.332	0.380	0.284	0.405	0.376	6
7	0.291	0.343	0.255	0.369	0.343	7
8	0.253	0.313	0.219	0.337	0.308	8

With $x =$ Aachen and $x =$ Potsdam, and with all five stations as y. In the special case $x = y$ we are faced with auto-correlation coefficients

6.3 Precipitation: Logistic Regression

We reduce the amount of daily precipitation ("Precip.") to the two alternatives "Precip. $= 0$" and "Precip. > 0". Accordingly, with the dichotomous variable Z, we have cases with $Z = 0$ and with $Z = 1$. The logistic regression approach models the probability π for $Z = 1$, see Appendix C.1.

Denoting by Y_t the precipitation amount at day t, we build the dichotomous variable Z_t. The probability for $Z_t = 1$ (that is for the event "Precip. > 0" at day t or "$Y_t > 0$") is abbreviated by

$$\pi_t = \mathbb{P}(Z_t = 1), \qquad t = 1, \ldots, n. \tag{6.2}$$

Using regressor variables, that are the lagged precipitation amounts with a time lag of one and two days, we form the *linear* regression terms

$$\eta_t(\alpha) = \alpha_0 + \alpha_1 \cdot Y_{t-1} + \alpha_2 \cdot Y_{t-2}, \qquad t = 1, \ldots, n, \tag{6.3}$$

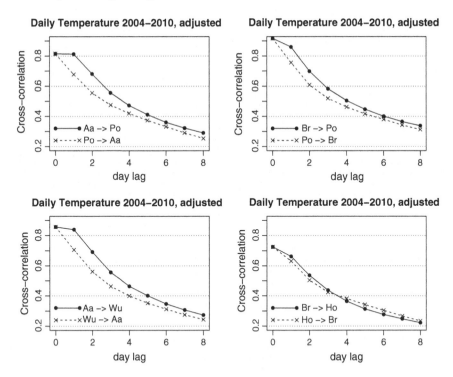

Fig. 6.4 Cross-correlation functions between stations, for daily temperature, 2004–2010. The series are seasonally adjusted. In each of the four *plots*, the two *curves* refer to the two directions "$x \to y$" and "$y \to x$". *Aa* Aachen, *Br* Bremen, *Ho* Hohenpeißenberg, *Po* Potsdam, *Wu* Würzburg

(Y_{-1}, Y_0 artificially). Here, $\alpha = (\alpha_0, \alpha_1, \alpha_2)$ is the vector of unknown parameters. With the η-terms from Eq. (6.3), the *logistic* regression model is constituted by the equation

$$\pi_t(\alpha) = \frac{\exp(\eta_t(\alpha))}{1 + \exp(\eta_t(\alpha))} = \frac{1}{1 + \exp(-\eta_t(\alpha))}, \qquad t = 1, \ldots, n. \qquad (6.4)$$

The parameters α_i are estimated by the maximum-likelihood method, cf. Appendix C.1. They are given in Table 6.3, together with the (negative) log-likelihood. All α_1 (and most α_2) values differ significantly from zero. The estimated coefficients α_i were used to calculate the *predicted probabilities* $\hat{\pi}_t$ according to (6.4). The calculation of $\hat{\pi}_t$ has to be based on the information up to time $t - 1$ (*forecast approach* of prediction). For this reason, we estimate the coefficients α_i for each time point t (greater than a starting value t_0) anew, as already done in 4.1 and 5.1 for yearly and monthly data. As a minimum sample for the estimation procedure we choose the days of the first year, i.e., we put $t_0 = 365$ (so that we have the $\hat{\pi}_t$ at our disposal only from t_0 onwards, that are for $n - t_0 = 2190$ cases). Further, on the basis of the predicted probabilities, we determine the number $P\%$ of *correctly classified* cases:

Table 6.3 Logistic regression for the daily precipitation amounts in the years 2004–2010, with lagged precipitation variables as regressors (truncated by 40 mm)

Station	Coefficients			Neg.			Predicting Obs		
	α_0	α_1	α_2	log-likeli	Median	$P\%$	$\hat{\pi}$	$\hat{Z}$	h
Aachen	−0.2440	0.1707	0.0219	1415.9	0.460	65.7%	0.448	0	10
Bremen	−0.3301	0.1891	0.0558	1423.4	0.446	64.8%	0.435	0	2
Hohenpeißenberg	−0.2118	0.1032	0.0041	1453.0	0.450	64.9%	0.447	0	9
Potsdam	−0.3768	0.1381	0.0571	1449.8	0.417	64.0%	0.419	1	1
Würzburg	−0.3523	0.1336	0.0429	1457.3	0.420	64.2%	0.416	0	2

Coefficients, neg. log-likelihood, median of predicted probabilities and percentage P of correctly classified cases are given. Further: the predicted probability $\hat{\pi}$ for the 1.1.2011 and the corresponding dichotomous variable $\hat{Z}$, together with the observed amount h (1/10 mm) at the 1.1.2011

Using the median med of all $\hat{\pi}_t$-values, we classify (predict) case t as Precip. $= 0$, if $\hat{\pi}_t \leq$ med, and as Precip. > 0, if $\hat{\pi}_t >$ med. Then we count, how often this prediction agrees with the actual observation Z_t. For Hohenpeißenberg for example, the result of this count is given in the following 2×2-table.

Observed	Predicted as		
	Prec $= 0$	Prec > 0	Σ
Prec $= 0$	693	367	1060
	0.654	0.346	1.0
Prec > 0	402	728	1130
	0.356	0.644	1.0
Σ	1095	1095	2190

We have $693 + 728 = 1421$ correctly predicted (classified) cases, that are $1421/2190*100 = 64.9\%$. That is to be compared with 50%, when coin tossing is applied, or—slightly better—with $1130/2190*100 = 51.6\%$, when one always predicts the more frequent alternative, that is here "Precip. > 0".

The last three columns of Table 6.3 deal with the day 1.1.2011. The predicted probability $\hat{\pi}$ is calculated for that day acc. to (6.4) and the corresponding $\hat{Z}$ value is derived (if $\hat{\pi} \leq (>)$ median, then $\hat{Z} = 0 (= 1)$). A comparison with the actual observed precipitation amount h (at the 1.1.2011) yields only one correct classification and leads to the suggestion, that the prediction approach is here more useful for assessing the goodness-of-fit of the model than for gaining practical forecasts for each single day.

Since the coefficients α_1, α_2 as well as the regressors assume non-negative values, the predicted probabilities for the event "Precip. > 0" exceed the bound

$$\frac{1}{1 + e^{-\alpha_0}} > 0.4 \quad \text{[for all five stations]},$$

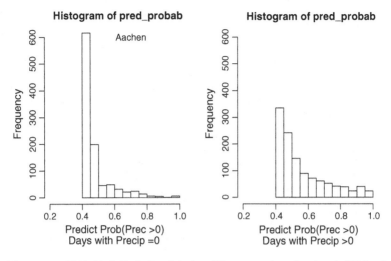

Fig. 6.5 Aachen, 2004–2010. Daily Precipitation. Histogram of predicted probabilities for the event "Precip. > 0", separately for cases without (Precip. = 0, *left*) and with (Precip. > 0, *right*) precipitation

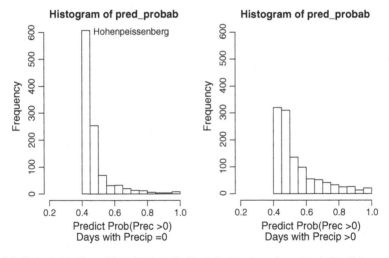

Fig. 6.6 Hohenpeißenberg, 2004–2010. Daily Precipitation. Same legend as in Fig. 6.5

see Figs. 6.5 and 6.6. The cases with "Precip. = 0" (left plot) have relatively more often low values (of the predicted probability) near the lower bound than the cases with "Precip. > 0" (right plot) have it.

Next, in Figs. 6.7 and 6.8, we plot the dichotomous observations Z_t and the predicted probabilities $\hat{\pi}_t$ (for the event "Precip. > 0") over the calendar days of the year. Due to the large oscillation of both quantities, we build centered (21-days) moving averages. In these plots, a familiar seasonal pattern cannot be discovered.

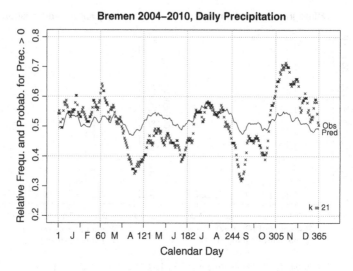

Fig. 6.7 Bremen, 2004–2010. Daily Precipitation. Course of observed frequencies (xxx) of the event "Precip. > 0" and of predicted probabilities (*solid line*) for the event "Precip. > 0", acc. to logistic regression, over the Calendar Days of the Year. Both *curves* are centered moving averages over $k = 21$ days, a second time averaged over the six years 2005–2010

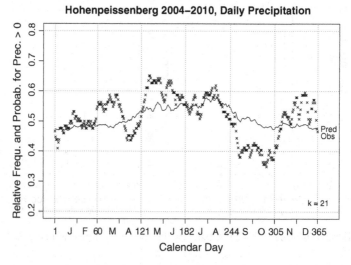

Fig. 6.8 Hohenpeißenberg, 2004–2010. Daily Precipitation. Same Legend as in Fig. 6.7

For all five stations, the observation curve sharply declines to a minimum in the months Apr., Sept., and Oct.: These are preferentially periods of dry days. We have such a minimum—in minor form—in the beginning of Jan., too. The curve of predictions accompanies that of observations more or less synchronously, better to see for Bremen (Fig. 6.7), worse for Hohenpeißenberg (Fig. 6.8).

R 6.1 Logistic Regression for daily precipitation data, by means of the R function glm. The criterion "amount of precipitation" (Prec) is divided into two categories (family=binomial), i.e., "Prec = 0" and "Prec > 0". Regressor variables are the lagged precipitation amounts, lag = 1 and lag = 2 (days). Further, histograms (hist) of the predicted probabilities for "Prec > 0" are established, see Fig. 6.5, and classification results w.r.to correctly classified cases are printed. An excerpt from the file Days5.txt, the daily temperature and precipitation data at five stations, can be found in Appendix A.3.

```
days<- read.table("C:/CLIM/Days5.txt",header=T)
attach(days)

postscript(file="C:/CLIM/Days.ps",height=6,width=16,horiz=F)
par(mfrow=c(1,2))

#--------------Data-----------------------------------------
quot<- "Aachen 2004-2010, Daily Precipitation"; quot
Pr<- pmin(PrAa/10,40)                     #Prec truncated by 40 [mm]
n<- length(Pr)

#--------------Logistic Regression--------------------------
"Logistic Regression, Two Alternatives"
"Predicted Prob (Prec >0)), Numerical (lagged) Variables PX,PY"
PX<- Pr; PX[3:n]<- Pr[1:(n-2)]            #PX[1:2] artificial
PY<- Pr; PY[2:n]<- Pr[1:(n-1)]           #PY[1]   artificial

tst<- 365;  ts1<- tst+1;  c("N Days"=n,"t start"=tst)

Pr01<- pmin(Pr*10,1)       #Numerical variable, alternatives 0,1
Q<- Pr; logli<- 0                         #Q vector of dim n
for(t in ts1:n){
prec.log<- glm(Pr01[1:t]~PY[1:t]+PX[1:t],family=binomial)
pred<- fitted(prec.log)                   #Predicted probabilities
Q[t]<- pred[t]                            #Forecast Approach
phat<- pmax(pmin(Q[t],0.999),0.001);  y<- Pr01[t]
logli<- logli + y*log(phat) + (1 - y)*log(1-phat)
}
c("LogLikelih_Forecast"=logli)
"Output for t=n"
summary(prec.log)

#--------------Histograms-----------------------------------
predt<- Q[ts1:n]; Prn<- Pr01[ts1:n]       #vectors of dim n-tst
pred0<- predt[Prn==0]; pred1<- predt[Prn==1]

#2 Histograms for predicted probabilities
xtext<- "Predict Prob(Prec >0)"
hist(pred0,xlim=c(0,1),ylim=c(0,600),xlab=xtext,ylab="Frequency")
title(sub="Days with Precip =0")
hist(pred1,xlim=c(0,1),ylim=c(0,600),xlab=xtext,ylab="Frequency")
```

```
title(sub="Days with Precip >0")

#--------------Classification-------------------------------
med<- median(predt)                    #Median of predicted probs
"Correctly classified Cases for t=ts1:n  [Forecast Approach]"
case00<-Prn[Prn==0 & predt<=med];case11<-Prn[Prn==1 & predt>med]
corrt<- length(case00) + length(case11) #Number of correct cases
casen<- length(pred0) + length(pred1)               #casen= n-tst
c("Median"=med,"Correct00"=length(case00),"from"=length(pred0),
              "Correct11"=length(case11),"from"=length(pred1))
c("Total"=corrt,"from"=casen,"Percent"=(corrt/casen)*100)

dev.off()
```

When augmenting the regression term (6.3) by further lagged precipitation variables (lags 3 and 4) or by lagged temperature variables, the goodness-of-fit of the model (log-likelihood, percentage of correct classification) is not essentially improved.

6.4 Precipitation: Categorical Data Analysis

This section offers contingency-table analysis, for the homogeneity problem (when comparing the five stations), and for the independence problem (when cross-correlating the stations).

Comparing the Five Stations by Contingency Tables

The amount of daily precipitation, measured in [mm] height, is divided into six categories, by dividing the interval $[0, \infty)$ into the six non-overlapping intervals

$$\frac{1 \quad 2 \quad 3 \quad 4 \quad 5 \quad 6}{[0] \ (0, 1.0] \ (1, 2.5] \ (2.5, 5] \ (5, 10] \ (10, \infty)}, \cdots \tag{6.5}$$

In Table 6.4 the number of cases, falling into the single intervals, is listed for each station, together with the relative frequencies (which add up to 1).

We are now going to test the hypothesis H_0 of *homogeneity*. H_0 asserts that the distribution of the precipitation amount over the six categories is identical for the five stations, see Appendix C.2. Under H_0 we expect the following frequencies and relative frequencies for each station.

Table 6.4 Daily precipitation amounts (mm) at five stations in the years 2004–2010, categorized

Station	m = 6 intervals						Sums
	[0]	(0, 1.0]	(1, 2.5]	(2.5, 5]	(5, 10]	(10, ∞)	
Aachen	1201	430	281	265	237	141	2555
	0.470	0.168	0.110	0.104	0.093	0.055	1
Bremen	1242	497	285	221	201	109	2555
	0.486	0.195	0.112	0.086	0.079	0.043	1
Hohenpeißenberg	1235	348	247	247	240	238	2555
	0.483	0.136	0.097	0.097	0.094	0.093	1
Potsdam	1324	486	233	227	193	92	2555
	0.518	0.190	0.091	0.089	0.076	0.036	1
Würzburg	1330	501	250	207	154	113	2555
	0.521	0.196	0.098	0.081	0.060	0.044	1
Sums	6332	2262	1296	1167	1025	693	12775
	0.496	0.177	0.101	0.091	0.080	0.054	1

The number of cases, falling into the single intervals, is given, together with the relative frequencies beneath

m = 6 intervals						Sums
[0]	(0, 1.0]	(1, 2.5]	(2.5, 5]	(5, 10]	(10, ∞)	
1266.4	452.4	259.2	233.4	205.0	138.6	2555
0.496	0.177	0.101	0.091	0.080	0.054	1

From Table 6.4 we derive Pearson's χ^2-test statistic $\hat{\chi}^2_{12775} = 187.1$. This value has to be compared with the quantile $\chi^2_{20, 0.99} = 37.57$ of the χ^2_{20}-distribution [DF $= (5 - 1) * (6 - 1) = 20$, $\alpha = 0.01$], such that the hypothesis H_0 of homogeneity is rejected. W.r.to the distribution of precipitation, there are significant differences between the five stations. Table 6.4 reveals that the first category [0] is especially frequent in Potsdam and Würzburg, the second category seldom and the last category frequent at Hohenpeißenberg, the third frequent in Aachen and Bremen.

In a further step of the analysis we ask, between which stations—in particular—the difference is significant. For this, we form all ten 2 × 6-subtables from Table 6.4 and calculate the test statistic $\hat{\chi}^2_{5110}$ with $1 * 5$ DF. We choose the subtable for the comparison (Potsdam, Würzburg) as an example.

Station	m = 6 intervals						Sums
	[0]	(0, 1.0]	(1, 2.5]	(2.5, 5]	(5, 10]	(10, ∞)	
Potsdam	1324	486	233	227	193	92	2555
	0.518	0.190	0.091	0.089	0.076	0.036	1
Würzburg	1330	501	250	207	154	113	2555
	0.521	0.196	0.098	0.081	0.060	0.044	1
Sums	2654	987	483	434	347	205	5110
	0.519	0.193	0.095	0.085	0.068	0.040	1

For this subtable we obtain the test statistic $\hat{\chi}^2_{5110} = 8.30$. Since 8.30 is smaller than the quantile $\chi^2_{5,0.999} = 20.52$, the hypothesis of homogeneity in the subtable is not rejected. The distributions of the precipitation amount for the two stations (Potsdam,Würzburg) do not differ significantly. Note, that we use the Bonferroni correction $1 - 0.01/10 = 0.999$, due to the 10 simultaneous pair comparisons.

Pearson's χ^2-test is applied to all 10 pair comparisons and is presented in the following symmetric scheme. Each value of the $\hat{\chi}^2_{5110}$-test statistic has to be compared with the quantile $\chi^2_{5,0.999} = 20.52$ and, if exceeding it, is marked by a double asterisk (**).

	Aachen	Bremen	Hohenp.	Potsdam	Würzburg
Aachen	–	16.60	36.78**	31.64**	41.63**
Bremen	16.60	–	81.86**	9.64	12.07
Hohenpeißenberg	36.78**	81.86**	–	96.88**	97.92**
Potsdam	31.64**	9.64	96.88**	–	8.30
Würzburg	41.63**	12.07	97.92**	8.30	–

Besides the comparison (Aachen, Potsdam) between the most western and the most eastern of the five stations, the comparison (Aachen, Würzburg) and all four comparisons with the mountain station Hohenpeißenberg turn out to be significant.

Cross-Correlation by Contingency Tables

The correlations of daily temperatures between the five stations were calculated—in Sect. 6.2—by means of the cross-correlation function from time series analysis. Since daily precipitation—with its frequent value 0—is no genuinely metric variable, we employ here once again contingency-table methods. This time we are dealing with the independence problem, see Appendix C.2. We divide the range $[0, \infty)$ of precipitation amount once again into the six intervals (6.5) introduced above. Let us denote the amount at station a and at day t, categorized into $1, \ldots, 6$, by Y_t^a.

Analogously, the amount at station b at day s is then Y_s^b. We will put $s = t + h$, with the time difference of h days.

Now we can form a 6×6 contingency table, where the entry n_{ij} of the table denotes the number of days t, where

$$Y_t^a = i \quad \text{and} \quad Y_{t+h}^b = j, \quad t = 1, \ldots, n - h; \ i, j = 1, \ldots, 6,$$

that is, where the amount at station a falls into category i and the amount at station b (h days later) into category j. As examples, we choose a = Aachen, b = Potsdam, $h = 1$ and a = Potsdam, b = Aachen, $h = 1$. Then we have the two contingency tables shown below. We will report Pearson's χ^2-test statistic $\hat{\chi}^2_m$ ($m = 2555 - h$) as well as Cramér's V,

$$V = \sqrt{\frac{\hat{\chi}_m^2}{m \cdot 5}}, \qquad 0 \le V \le 1 ;$$

V serves us as a substitute for the correlation coefficient.

Y_t^a	a = Aachen → b = Potsdam Y_{t+1}^b						
	[0]	(0, 1.0]	(1, 2.5]	(2.5, 5]	(5, 10]	(10, ∞)	Sums
[0]	825	180	67	60	48	21	1201
(0, 1.0]	191	94	42	43	36	23	429
(1, 2.5]	111	65	38	32	23	12	281
(2.5, 5]	86	70	42	31	29	7	265
(5, 10]	71	54	25	35	36	16	237
(10, ∞)	39	23	19	26	21	13	141

We obtain Pearson's test statistic $\hat{\chi}_{2554}^2 = 336.7$ with 25 degrees of freedom; one calculates $V = \sqrt{336.7/(5 * 2554)} = 0.162$.

Y_t^a	a = Potsdam → b = Aachen Y_{t+1}^b						
	[0]	(0, 1.0]	(1, 2.5]	(2.5, 5]	(5, 10]	(10, ∞)	Sums
[0]	768	185	112	110	92	57	1324
(0, 1.0]	197	92	61	57	49	29	485
(1, 2.5]	89	49	37	23	26	9	233
(2.5, 5]	76	47	31	25	30	18	227
(5, 10]	51	37	26	36	26	17	193
(10, ∞)	20	19	14	14	14	11	92

We obtain the test statistic $\hat{\chi}_{2554}^2 = 171.6$ with 25 DF, from where one derives $V = \sqrt{171.6/(5 * 2554)} = 0.116$.

Letting h run from 0 to 8 days, we arrive at the following two lists of V-values, which are visualized in Fig. 6.9, second plot.

Lag (days)	0	1	2	3	4	5	6	7	8
Aa → Po	0.196	0.162	0.101	0.072	0.058	0.043	0.055	0.051	0.038
Po → Aa	0.196	0.116	0.075	0.070	0.057	0.043	0.039	0.050	0.042

We observe here the same phenomenon, which we have already noticed with daily temperature data (in Fig. 6.4). The cross-correlation curve for the west-east direction (the main wind direction) lies above that for the reversed direction, up to a time lag of 2 or 4 days (Fig. 6.9). In the case of Bremen-Würzburg, there is no clear preference for one of the two curves.

R 6.2 Categorical Data Analysis for daily precipitation. Cross-correlation between stations (up to a time lag of 8 days) by contingency tables (`table`), Pearson's χ^2 statistic and Cramér's V. This is done by `chisq.test` and by the user functions `Vabba`, `Vau`, `plotCrosV`. The latter produces a plot for two stations A and B and for cross-correlations A → B and B → A, to be seen in Fig. 6.9. Instead of `chisq.test` the user function `ChiSqu` in the supplement can also be used.

The precipitation amount is divided into 6 categories by the R command `cut`.

```
days<- read.table("C:/CLIM/Days5.txt",header=T)
attach(days)
postscript(file="C:/CLIM/DaysCros.ps",height=6,width=16,horiz=F)
#par(mfrow=c(2,2))                                   #for further plots


#----------------------------------------------------------------
Vau<- function(Xa,Xb,m,n,l){    #Cramer's V of an mxm table YaxYb
Ya<- Xa[1:(n-l)]; Yb<- Xb[(1+l):n]                   #time lag l
chi2<-chisq.test(Ya,Yb)$statistic   #Pearson's $chi^2 statistic
#or with the user function ChiSqu:
#tab<- table(Ya,Yb); chi2<- ChiSqu(tab,m,m)
Vau<- sqrt(chi2/((n-l)*(m-1)))
return(Vau)
}


Vabba<- function(Pra,Prb,m,n,brk,lx){           #cross corr fuction
Preca<- pmin(Pra/10,100); Precb<- pmin(Prb/10,100)
Prca<-  cut(Preca,brk)                    #Categorical variable Prca
Prcb<-  cut(Precb,brk)                    #Categorical variable Prcb
Vab<- 1:(lx+1); Vba<- 1:(lx+1)
for(l in 0:lx){
Vab[l+1]<- Vau(Prca,Prcb,m,n,l)
Vba[l+1]<- Vau(Prcb,Prca,m,n,l) }
return(cbind(Vab,Vba))           #cbind produces an (lx+1)x2 matrix
}


plotCrosV<- function(mx,cra,crb,yc,xtxt,ytxt,ttxt,Stab,Stba){
plot(0:mx,cra,type="l",lty=1,xlim=c(0,mx),ylim=yc,
            xlab="day lag",ylab="Cramers V")
points(0:mx,cra,pch=16);   title(main=ttxt,cex=0.6)
lines(0:mx,crb,lty=2);   points(0:mx,crb,pch=4)
legend(xtxt,ytxt,legend=c(Stab,Stba),lty=c(1,2))
}


#----------------------------------------------------------------
ttxt<- "Daily Precipitation 2004--2010"; ttxt
n<- length(Year)
m<- 6; brk<-c(-1,0,1,2.5,5,10,100)                      #6 categories
"6 Classes: [0],(0,1],(1,2.5],(2.5,5],(5,10],(10,100])"
yc<- c(0.0,0.26); xtxt<- 5; ytxt<- 0.25     #plotting parameters
lx<- 8                                       #maximal time lag
```

```
Pra<- PrAa; Prb<- PrHo
Stab<- "Aa -> Ho"; Stba<- "Ho -> Aa"
V<- Vabba(Pra,Prb,m,n,brk,lx)
Vab<- V[,1]; Vba<- V[,2]              #1. and 2. column of matrix V
Stab; Vab; Stba; Vba
plotCrosV(lx,Vab,Vba,yc,xtxt,ytxt,ttxt,Stab,Stba)

#Continue with: Pra<- PrAa; Prb<- PrPo   etc.
dev.off()
```

Supplement

```
ChiSqu<- function(mat,k,m)  #Pearson's chi^2 of a kxm matrix mat
{n<- sum(mat); ch2<- 0
for(i in 1:k){ni<- sum(mat[i,])                          #row sums
for(j in 1:m){nj<- sum(mat[,j])                       #column sums
eij<- ni*nj/n                              #expected frequencies
ch2<- ch2+(mat[i,j] - eij)^2/eij}}
return(ch2)
}
```

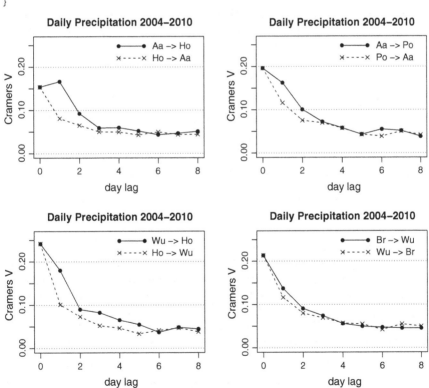

Fig. 6.9 Cross-correlations between stations by means of Cramér's V for contingency tables. Daily Precipitation, 2004–2010. In each of the four *plots*, the two *curves* relate to the two directions "$x \to y$" and "$y \to x$". *Aa* Aachen, *Br* Bremen, *Ho* Hohenpeißenberg, *Po* Potsdam, *Wu* Würzburg

6.5 Heavy Precipitation: Event-Time Analysis

In event-time analysis the focus of attention lies on the occurrences of a certain kind of event; in the following it will be the occurrences of daily precipitation heights above a predefined bound B (mm). We write down the time points (here: days) t_1, $t_2, \ldots, t_n$ when such an event occurs.

In what follows, it is the sample $t_1, t_2, \ldots, t_n$ of n event-times (also called *occurrence times*) which is analyzed. Our main tools of analysis will be intensity functions and (inhomogeneous) Poisson processes.

Counting Processes and Intensity Processes

The basic object in event-time analysis is a non-negative random function $\lambda(t)$, $t \geq 0$, called *intensity* function (or intensity process). For $\lambda(t)$ we presuppose (a certain kind of) continuity, and for the integrated intensity function

$$\Lambda(t) = \int_0^t \lambda(s)\, \mathrm{d}s$$

we assume $\Lambda(t) < \infty$ for all $t \geq 0$. We further define the *counting* process N_t, $t \geq 0$, by the number of events, occurring in the time interval $[0, t]$. For $s \leq t$, the *increment* $N_t - N_s$ is the number of events in the interval $(s, t]$. The relation between the (observable) counting process and the (unknown) intensity process is

$$\lambda(t) = \lim_{h \to 0} \frac{1}{h}\, \mathbb{E}(N_{t+h} - N_t | \mathcal{F}_t) = \lim_{h \to 0} \frac{1}{h}\, \mathbb{P}(N_{t+h} - N_t \geq 1 | \mathcal{F}_t), \qquad (6.6)$$

$\mathcal{F}_t$ being a theoretical concept of the "information before time t". The second equation of (6.6) is proven in counting process theory. An interpretation: $\lambda(t)$ gives the tendency that an event occurs around time point t (similar to the concept of a density function in probability theory; but the intensity function is not normalized to have integral 1). Further the equation $\mathbb{E}(N_t) = \mathbb{E}(\Lambda(t))$ is valid.

Guided by Eq. (6.6), we can derive a nonparametric *curve estimator* for $\lambda(t)$ on the basis of a sample $t_1, t_2, \ldots, t_n$ by

$$\hat{\lambda}(t) = \frac{1}{h} \cdot \sum_{i=1}^{n} K\left(\frac{t - t_i}{h}\right), \qquad (6.7)$$

where K is a *kernel* and h is a positive band (window) width; see Andersen et al. (1993), II 4.1, IV 2.1.

Two examples: The Gaussian kernel and the rectangular kernel, respectively, are

$$K(s) = \frac{1}{\sqrt{2\pi}} \exp\left(-\frac{1}{2}s^2\right), \quad \hat{\lambda}(t) = \frac{1}{\sqrt{2\pi}h} \sum_{i=1}^{n} \exp\left(-\frac{(t-t_i)^2}{2h^2}\right)$$

$$K(s) = \begin{cases} 1/2 & |s| \leq 1 \\ 0 & \text{else} \end{cases}, \quad \hat{\lambda}(t) = \frac{1}{2h} \sharp\{i : t-h \leq t_i \leq t+h\},$$

the latter being an empirical counterpart to (6.6).

Poisson Processes

An *inhomogeneous* Poisson process has a deterministic intensity function $\lambda(t)$ and the properties

1. For all $s \leq t$, the increment $N_t - N_s$ is independent of the realization $(N_u, u \leq s)$ of the process up to time s.
2. For all $s \leq t$, the increment $N_t - N_s$ is Poisson distributed with parameter $\Lambda(s, t) = \Lambda(t) - \Lambda(s) = \int_s^t \lambda(u)\mathrm{d}u$, i.e.,

$$\mathbb{P}(N_t - N_s = k) = \frac{(\Lambda(s, t))^k}{k!} \cdot \exp(-\Lambda(s, t)).$$

3. For each k, the waiting time $S_k = T_{k+1} - T_k$ till the next event has (conditionally) an exponential distribution, i.e.,

$$Q_k(s) = \mathbb{P}(S_k \leq s | (T_1, \ldots, T_k)) = 1 - \exp(-\Lambda(T_k, T_k + s)).$$

Hereby, we interpret the sample $(t_1, \ldots, t_n)$ as a realization of random variables $(T_1, \ldots, T_n)$. See Cox and Lewis (1966) or Snyder (1975) for more information and for important applications.

As likelihood of a sample $(t_1, t_2, \ldots, t_n)$ of n occurrence times within a (predefined) time interval $[0, t_b]$, we write down

$$f(t_1, \ldots, t_n) = \left(\prod_{i=1}^{n} g(t_{i-1}, t_i)\right) \cdot \exp(-\Lambda(t_n, t_b)). \tag{6.8}$$

Hereby we have set $t_0 = 0$ and

$$g(t_{i-1}, t_i) = \lambda(t_i) \cdot \exp\left(-\Lambda(t_{i-1}, t_i)\right).$$

The last factor in (6.8) is the probability that there is no event between t_n and t_b. Eq. (6.8) amounts to

$$f(t_1, \ldots, t_n) = \left(\prod_{i=1}^{n} \lambda(t_i) \right) \cdot \exp(-\Lambda(t_b)), \quad \Lambda(t_b) = \int_0^{t_b} \lambda(s) ds,$$

from where we obtain the log-likelihood function

$$\ell_n = \log f(t_1, \ldots, t_n) = \sum_{i=1}^{n} \log \lambda(t_i) - \Lambda(t_b); \qquad (6.9)$$

see also Andersen et al. (1993), II 7.2. We call the process a *homogeneous* Poisson process, if the intensity function is constant over time: $\lambda(t) = \lambda$ for all t. Here, we have in property 3

$$Q_k(s) = Q(s) = 1 - \exp(-\lambda \cdot s),$$

such that in this special case the waiting time $S_k = T_{k+1} - T_k$ is independent of the last event time T_k. The log-likelihood function (6.9) becomes $\ell_n = n \cdot \log \lambda - t_b \cdot \lambda$.

Statistics and Application to Daily Precipitation Data

As announced above, the time index $t = 1, 2, \ldots$ counts the successive days, from the starting point $t = 1$, that is the 1st January 2004, onwards, up to $t = t_b = 2555$, that is the 31st December 2010 (29th February canceled). The events of interest are daily precipitation heights above $B = 15\,\text{mm}$ at the stations Aachen, Bremen, Potsdam, Würzburg, and above $B = 20\,\text{mm}$ at Hohenpeißenberg. With a sample of n events, at the days $t_1, t_2, \ldots, t_n$, the above formulas (6.7) and (6.9) are applied. First, we provide nonparametric curve estimators (6.7) for the unknown intensity function $\lambda(t)$, using a Gaussian kernel with band width $h = 40$ days. Figures 6.10 and 6.11 (*top*) present the estimated intensity curves for Hohenpeißenberg and Würzburg, demonstrating the existence of a strong yearly periodicity (seasonality) and of a weaker trend. Therefore, when establishing a parametric intensity, we incorporate into the function a (quadratic) trend term and a (sin/cos) seasonal term. With five parameters

$$\theta = (a, b_1, b_2, c, d)$$

we define the intensity function $\lambda(t)$ for the *full* model by

$$\lambda(\theta, t) = \exp\left(a + m((b_1, b_2), t) + s((c, d), t) \right),$$
$$m((b_1, b_2), t) = m(t) = b_1 \cdot t + b_2 \cdot t^2, \qquad (6.10)$$
$$s((c, d), t) = s(t) = c \cdot \sin(\omega t) + d \cdot \cos(\omega t),$$

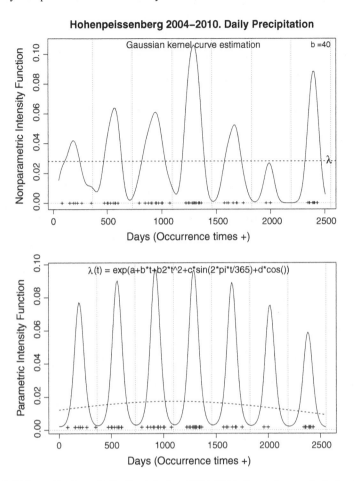

Fig. 6.10 Hohenpeißenberg, Daily Precipitation, 2004–2010. Event-time analysis for days with precipitation amount > 20 mm. *Top* Nonparametric curve estimation of the intensity function acc. to Eq. (6.7), by using a Gaussian kernel and a band width $b = 40$ days. *Bottom* Parametric estimation of the intensity function for the full model (*solid line*) acc. to (Eq. 6.10) and for the submodel with quadratic trend term only (dashed line) acc. to (Eq. 6.11)

where $\omega = (2 \cdot \pi)/365$. Further, for testing purposes, we also apply *sub*-models possessing the intensity functions

$$\lambda_a(a, t) = \lambda_a(t) = \exp(a) \quad \text{[constant intensity model]},$$
$$\lambda_m((a, b_1, b_2), t) = \lambda_m(t) = \exp\left(a + m((b_1, b_2), t)\right) \quad \text{[trend model]}, \quad (6.11)$$
$$\lambda_s((a, c, d), t) = \lambda_s(t) = \exp\left(a + s((c, d), t)\right) \quad \text{[seasonal model]}.$$

The constant intensity model belongs to a homogeneous Poisson process. The use of the exponential function exp guarantees us the positivity of the intensities.

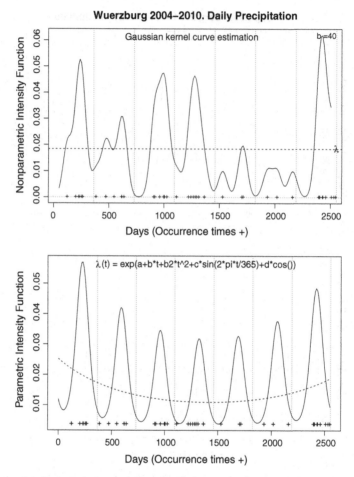

Fig. 6.11 Würzburg, Daily Precipitation, 2004–2010. Event-time analysis for days with precipitation amount >15 mm. Legend as in Fig. 6.10

According to the four intensity functions $\lambda_a, \lambda_m, \lambda_s, \lambda$, presented above, we have four log-likelihood functions

$$\ell_a \,, \; \ell_m \,, \; \ell_s \,, \; \ell$$

for the homogeneous (constant intensity) model, the trend model, the seasonal model [see Eq. (6.11)] and for the full model (6.10), respectively. The unknown parameters θ are estimated by maximizing the log-likelihood function (6.9), resulting in an ML-estimator for θ. This is done numerically by a grid-search method and by numerical integration to get $\Lambda(t_b)$. In the homogeneous (constant intensity) model, the ML-estimator of the constant a is

$$a \;=\; \log(\hat{\lambda}) \;=\; \log(n/t_b) \,.$$

Table 6.5 Event-time analysis for days with precipitation amount above B mm, at the stations Aachen, Bremen, Hohenpeißenberg, Potsdam, Würzburg, 2004–2010

Sta.	B	n	$\ln \hat{\lambda}$	$(a, b_1 \cdot 10^3, b_2 \cdot 10^7, c, d)$	T(a:m)	T(a:s)	T(m)	T(s)
Aa	15	70	−3.597	(−3.61, 0.11, −0.83, −0.32, −0.33)	0.30	6.75	6.91	0.46
Br	15	55	−3.838	(−4.49, 0.83, −2.82, −0.60, −0.65)	1.49	18.98	18.62	1.13
Ho	20	72	−3.569	(−4.41, 0.66, −3.02, −0.29, −1.70)	1.79	68.74	68.69	1.74
Po	15	41	−4.132	(−5.12, 1.42, −4.74, −0.40, −0.73)	2.66	12.88	12.61	2.39
Wu	15	47	−3.996	(−3.68, −1.18, 4.18, −0.76, −0.75)	1.70	21.36	21.89	2.23

The parameters θ of the full model $\lambda(\theta, t)$ acc. to Eq. (6.10) and the log-LR test statistics (6.12), (6.13) to test the submodels are presented

By means of the log-likelihood ratio (log-LR) test statistics

$$T(a : m) = 2 \cdot (\ell_m - \ell_a), \quad T(a : s) = 2 \cdot (\ell_s - \ell_a), \tag{6.12}$$

we test the hypotheses H_0 of a homogeneous (constant intensity) model within the larger models with quadratic trend and with seasonality, respectively. Under H_0, they are χ^2-distributed with 2 DF (for larger n).

By means of the log-LR test statistics

$$T(m) = 2 \cdot (\ell - \ell_m), \quad T(s) = 2 \cdot (\ell - \ell_s), \tag{6.13}$$

we test the hypotheses H_0 of a quadratic trend model and of a seasonal model, respectively, within the full model (6.10). Under H_0, they are χ^2-distributed with 2 DF (for larger n). The $\chi^2_{2,1-\alpha}$-quantiles are

$$\alpha = 0.10 : 4.605, \quad \alpha = 0.05 : 5.992, \quad \alpha = 0.01 : 9.210.$$

Table 6.5 reports the four test statistics for the five stations. The extension of the constant intensity by a *quadratic term* is not significant, neither is it the extension of the sin/cos term to the full term by the quadratic term, according to test statistics T(a:m) and T(s), respectively. This is different from the *sin/cos term*, which forms a significant extension of the constant intensity (see T(a:s)) and significantly extends the quadratic term to the full term (see T(m)). The significance level is $\alpha = 0.01$ (Aachen: 0.05). Although not so evident from the lower plots in Figs. 6.1, 6.2, 6.3, the yearly periodicity (seasonality) is very strong (and dominates the trend) in the series of days with heavy precipitation.

Figures 6.10 and 6.11 (*bottom*) show the intensity functions $\lambda(\theta, t)$ for the full model, according to Eq. (6.10), and for the submodel with the quadratic trend term only, i. e. $\lambda_m(t) = \lambda_m((a, b_1, b_2), t)$ cf. Eq. (6.11), plotted over the seven years 2004–2010. At the stations Aachen, Bremen, Potsdam (no Figs.) and Hohenpeißenberg the intensity curves $\lambda_m(t)$ are concave, with a decrease in the last 2 or 3 years. That is different from Würzburg, where the curve is convex (with a positive coefficient b_2 and an increase in the last 3 years).

R 6.3 Event-time analysis for days with precipitation amount above *B* mm: Non-parametric estimation of the intensity function by kernel method and plot of the resulting curve estimator, as in Figs. 6.10 and 6.11 (top). The choice is between the Gaussian and the rectangular kernel (user functions gauss and rectang, resp.) In the plot a margin is provided by means of the logical vector red of dimension ndelt, taking here the values FALSE FALSE FALSE TRUE ... TRUE FALSE FALSE FALSE FALSE. Data are read from file C:/CLIM/eventAa.txt in the form of a vector $(n, tx(1), \ldots, tx(n))$; tx being the occurrence times of the n events in Aachen in the years 2004–2010. We have $B = 15$, $n = 70$ and tx is the vector

```
  12   19   127   190   203   222   225   265   266   278   322   351   385   406
 408   480   591   616   623   660   845   870   876   877   946   956   963  1053
1075  1112  1113  1254  1256  1264  1280  1315  1316  1328  1365  1371  1397  1409
1435  1539  1555  1564  1615  1623  1643  1651  1669  1675  1715  1733  1738  1866
1872  1932  1987  2049  2108  2316  2321  2336  2417  2428  2506  2507  2543  2547
```

```
postscript(file="C:/CLIM/Intfunc.ps",height=6,width=16,horiz=F)

#----------------------------------------------------------------
gauss<- function(t,x,b){                                #Gauss kernel
c<- sqrt(2*pi)
fun<- (1/c)*exp(-(t-x)^2/(2*b*b))
return(fun)
}
rectang<- function(t,x,b){                              #Rectangular kernel
fun<- 0
if({t-b <= x} & {x<= t+b}) fun<- 1/2
return(fun)
}


#----------------------------------------------------------------
quot<- "Aachen 2004--2010. Daily Precipitation"; quot

xx<- scan("C:/CLIM/eventAa.txt")
n<- xx[1];   tx<- xx[2:(n+1)];   tb<- 2555
c("n"=n,"Right end"=tb,"lambda"=n/tb,"log lambda"=log(n/tb))

"Nonparametric kernel estimation"
#Curve evaluated at ndelt points
ndelt<-200; delt<- tb/ndelt; bh<- 40                    #bh bandwidth
tt<- delt*(1:ndelt); int<- 1:ndelt              #vectors of dim ndelt

for(j in 1:ndelt){ kern<- 0
for(i in 1:n){                          #choose:gauss( ) or rectang( ):
kern<- kern + gauss(tt[j],tx[i],bh)}
int[j]<- (1/bh)*kern}

d<- bh; red<- tt>d & tt<tb-d     #logical vector, margin width d
ttr<- tt[red]; intr<- int[red]
```

```
c("ndelt"=ndelt,"delta"=delt,"bandwith"=bh,
                       "mean int"=mean(intr))

plot(ttr,intr,type="l",lty=1,xlab="Days (Occurrence times +)",
                     ylab="Nonparametric Intensity Function")
title(main=quot)
abline(h=n/tb,lty=2)
text(tb+10,n/tb,"l",font=5)                          #Greek lambda
abline(v=365*(1:7),lty=3)
text(tx,min(intr),"+",cex=0.8)              #mark occurrence times

dev.off()
```

Output from R 6.3 Notice: We have `mean int` $\approx \lambda = \frac{70}{2555}$.

"Aachen 2004--2010. Daily Precipitation".

n	Right end	lambda	log lambda
70.0	2555.0	0.02739	-3.59731

"Nonparametric kernel estimation"

ndelt	delta	bandwith	mean int
200.0	12.7750	40.0	0.02672

Chapter 7
Spectral Analysis

The analysis in the frequency domain, which now follows, is guided by the idea, that the "oscillation" of the observed series is produced by an overlapping of periodic (sin, cos) functions.

To detect "hidden" periodicities, we employ periodograms and smoothed periodograms, considered as estimators for the spectral density of the time series, see Appendix B.2. The problems with periodograms in the presence of a trend and with periodograms after the removal of the trend are discussed. Further, simultaneous statistical bounds are used to assess peaks of the (smoothed) periodograms. Wavelet analysis is able to trace periodicities which vary in the course of the ongoing process. So if the periodogram has large values at several period lengths, then it is sometimes possible to allocate them to different parts of the underlying time interval.

7.1 Periodogram, Raw and Smoothed

To describe periodical phenomena in time series Y_t, $t = 1, \ldots, n$, we employ—equivalently—

- the number k, $k = 1, \ldots, n/2$; i.e. the number of cycles in the time interval $[0, n]$,
- the period length $T = T_k = n/k$,
- the angular frequency $\omega = \omega_k = (2\pi/n) \cdot k$.

The *periodogram* is calculated by using the n Fourier coefficients of the time series. These coefficients

$$a_0, a_1, \ldots, a_{n/2}, \quad b_1, \ldots, b_{n/2-1}$$

H. Pruscha, *Statistical Analysis of Climate Series*,
DOI: 10.1007/978-3-642-32084-2_7, © Springer-Verlag Berlin Heidelberg 2013

(n supposed to be even) are gained by the formulas

$$a_0 = \frac{2}{n} \sum_{t=1}^{n} Y_t = 2\bar{Y}, \qquad a_{n/2} = \frac{1}{n} \sum_{t=1}^{n} (-1)^t Y_t ,$$

$$a_k = \frac{2}{n} \sum_{t=1}^{n} Y_t \cos(\omega_k t), \quad k = 1, \ldots, n/2 - 1, \tag{7.1}$$

$$b_k = \frac{2}{n} \sum_{t=1}^{n} Y_t \sin(\omega_k t), \quad k = 1, \ldots, n/2 - 1.$$

We define the *periodogram* $I(\omega_k), k = 1, 2, \ldots, n/2$, by using the sum of the squared Fourier coefficients a_k, b_k, more precisely by the equations

$$I(\omega_k) = \frac{n}{4\pi} \cdot (a_k^2 + b_k^2), \qquad\qquad k = 1, 2, \ldots, n/2 - 1,$$

$$I(\omega_{n/2}) = I(\pi) = \frac{n}{\pi} \cdot a_{n/2}^2 . \tag{7.2}$$

It is plotted over $k = 1, 2, \ldots, n/2$, respectively over $T = n, n/2, \ldots, 2$. It informs us, how strong a cycle with number k is involved in the oscillation of the time series. The plot of the periodogram generally looks very "jagged". By smoothing the periodogram we arrive at an estimator $\hat{f}(\omega)$ for the *spectral density* $f(\omega)$ of the time series; see the Appendix B.2.

A simple method of smoothing is the application of the so-called discrete *Daniel window*: moving averages are built over $2M + 1$ values of the periodogram, M values left and M values right of ω_k, leading to a special version of the spectral density estimator, namely to

$$\hat{f}(\omega_k) = \frac{1}{2M + 1} \cdot \big(I(\omega_{k-M}) + \cdots + I(\omega_k) + \cdots + I(\omega_{k+M}) \big).$$

We worked with $M = 5$, that is with $2M + 1 = 11$ points, throughout.

The periodogram $I(\omega)$—and hence $\hat{f}(\omega)$ too—is often *standardized* in the sense, that we divide it by s^2, the empirical variance of the time series.

7.2 Statistical Bounds

We start with reporting a central result. The ratios

$$\text{(i)} \quad \frac{I(\omega)}{f(\omega)} \quad \text{and} \quad \text{(ii)} \quad \nu \cdot \frac{\hat{f}(\omega)}{f(\omega)}, \qquad \omega = \frac{2 \cdot \pi}{T},$$

where $f(\omega)$ is the (true) spectral density of the time series, have asymptotically

(i)   an exponential distribution (with parameter 1) and
(ii) a χ_ν^2-distribution (with ν degrees of freedom),

respectively. It is $\nu = 4 \cdot M + 2$ in the case of the discrete Daniel window, which we are using here.

Recall, that we have $f(\omega) = \sigma^2/\pi$ for a white noise process, also called pure random series. From there we derive simultaneous bounds b_l and B_l,

$$b_l = -\frac{1}{\pi} \cdot \ln\left(\frac{\alpha}{l}\right), \qquad B_l = \frac{1}{\pi \nu} \cdot \chi_{\nu,1-\alpha/l}^2, \qquad \nu = 4M + 2, \qquad (7.3)$$

see Brockwell and Davis (2006, 10.3–10.5). They refer to the standardized periodogram (the b_l's) and spectral density estimation (the B_l's) of a pure random series, with the same variance as the observed time series Y_t. In Eq. (7.3) we have denoted the γ-quantile of the χ^2-distribution with ν degrees of freedom by $\chi_{\nu,\gamma}^2$. The meaning of these bounds is the following (for the standardized periodogram as example; in what follows we often suppress the attribute "standardized"). The probability that the maximum of l periodogram values (at l points ω_k resp. T_k, fixed in advance) of a pure random series exceeds the bound b_l, approximately amounts to α (here $\alpha = 0.05$). The *Bonferroni-correction* in (7.3), that is α/l instead of α, refers to the fact, that we base a rejection of the hypothesis of a pure random series not on the periodogram value at one single point, but on the values at several (namely l) points. In any case, the individual $l = 1$-bound is too low: note, that 5 % of the periodogram values of a pure random series lies—on the average—above the bound b_1.

Somewhat arbitrarily, we will speak of weak significance (of significance) of a periodogram value or of a smoothed periodogram value, if the b_4 or B_4 bound (the b_{12} or B_{12} bound) is exceeded. In the following figures, these bounds b_l and B_l, $l = 1, 4, 12$, are drawn as horizontal lines.

AR(1)-Correction. Trend Removal

If we take the auto-correlation $r = r(1)$ of the time series into account, we correct these bounds by a factor $\lambda(\omega)$. More precisely: under the assumption of an AR(1)-process, we have to multiply b_l and B_l by

$$\lambda(\omega) = \frac{1 - r^2}{1 - 2r \cos\omega + r^2}, \qquad \omega = \frac{2 \cdot \pi}{T}, \qquad (7.4)$$

arriving at the AR(1)-adjusted bounds

$$b_l(\omega) = b_l \cdot \lambda(\omega) \qquad \text{and} \qquad B_l(\omega) = B_l \cdot \lambda(\omega).$$

To give an argument: $\frac{s^2}{\pi} \cdot \lambda(\omega)$ is an estimator of the spectral density $f(\omega)$ of an AR(1)-process; see Eq. (B.7). This AR(1)-correction declares high spectral values, which are due to the AR(1) structure only [see the figure beside Eq. (B.7)], as non-significant. In

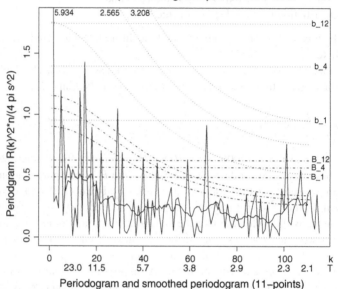

Fig. 7.1 Hohenpeißenberg. Annual temperature means. The standardized periodogram (*zigzag line*) and smoothed periodogram (*inner solid line*) of the time series. The bounds b_1, b_4, b_{12} for the periodogram (...) and B_1, B_4, B_{12} for the smoothed periodogram (–·–·–) are drawn as horizontal lines. The corresponding AR(1)-adjusted bounds $b_1(\omega)$, $b_4(\omega)$, $b_{12}(\omega)$ (...) enter the plot as S-type curves in increasing order (no labels). The same is the case for the AR(1)-adjusted bounds $B_1(\omega)$, $B_4(\omega)$, $B_{12}(\omega)$ (–·–·–). For the periodogram and for the curves $b_4(\omega)$ and $b_{12}(\omega)$, the truncated values at $k = 1$ can be found at the *upper border*

the case of $r(1) > 0$, these are spectral values for small ω- (large T-) values. Further in this case, spectral values for large ω- (small T-) values may become significant by such a correction. See also Schönwiese (2006, 14.6) in connection with climate applications.

Starting with Sect. 7.3, we do not analyze the observed series, but the series of *residuals* from a polynomial trend (here polynomials of order four were employed). *Without* this trend removal long-term fluctuations ($T \geq 20$ years or more) may dominate the periodogram- or spectral density plot, *with* trend removal they enter the plots in a weakened form only. Of course, the removal of a trend component may also remove true periodicities from the series.

The $r = r(1)$-values of the climate series after *trend adjustment* are very small—according to Tables 3.1 and 3.2, such that the correction term $\lambda(\omega)$ in Eq. (7.4) is approximately 1. Therefore, in detrended series, the AR(1)-correction will not be performed.

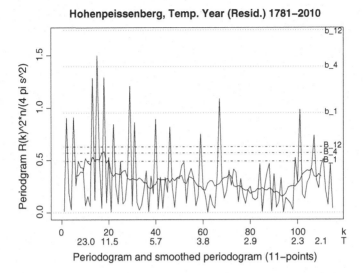

Fig. 7.2 Hohenpeißenberg. Annual temperature means (detrended). The standardized periodogram (*zigzag line*) and the standardized smoothed periodogram (*inner solid line*) of the time series of residuals from polynomial(4)-trend. The bounds b_1, b_4, b_{12} for the periodogram (…) and B_1, B_4, B_{12} for the smoothed periodogram (– ·– ·– ·–) are drawn as *horizontal lines*. The periodogram shows peaks ($>b_1$) at the periods of $T = 17.7$, 15.3, 12.8, 7.9, 3.4, 2.3 (years)

Yearly Temperature

In Fig. 7.1 we find periodogram and smoothed periodogram (i.e. spectral density estimation with discrete Daniel window) for annual temperature means at Hohen-peißenberg. With respect to the horizontal lines b_4 and b_{12}, we have the significant period of $T = 230$ years ($k = 1$) and the weakly significant period $T = 15.3$ ($k = 15$). The peak at this period is very sharp and small, so that here the smoothed periodogram does not exceed the horizontal bound B_4. Now we have an auto-correlation $r = r(1) = 0.29$, which is distinctly different from zero. Therefore, we correct the bounds b_l and B_l by the factor $\lambda(\omega)$ as in Eq. (7.4) and obtain the S-type curves $b_l(\omega)$ and $B_l(\omega)$ of the figure. The period $T = 230$ is still significant, but no other peak exceeds $b_4(\omega)$. Looking at the smoothed periodogram, it is now the period of $T = 2.2$ years, where the curve reaches the bound $B_4(\omega)$ and shows therefore weak significance.

Detrended series. Periodogram analysis of the series *after* the removal of the polynomial(4)-trend is shown in Fig. 7.2. The significant peak at $T = 230$ from Fig. 7.1 disappears and—instead—we have non-significant periodogram peaks at periods $T = 115$ and 46. We can state that in Fig. 7.1 the whole trend component is interpreted as one long cycle of $T = 230$ years. Notice that the rest of the peri-odogram keeps more or less unaltered when the trend component is removed.

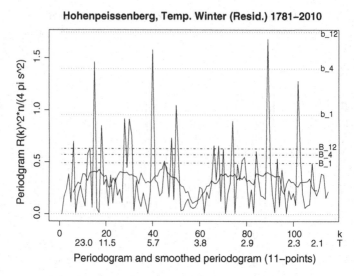

Fig. 7.3 Hohenpeißenberg. Winter temperature means (detrended). Legend as in Fig. 7.2. The periodogram shows peaks ($>b_1$) at the periods of $T = 15.3, 5.8, 4.6, 2.6, 2.3$ (years)

7.3   Yearly and Winter Temperature, Detrended

We remind that from now on the time series are trend-adjusted in the sense, that the residuals from a polynomial(4)-trend are built. Further: periodogram values above the horizontal line b_4 are called weakly significant, above b_{12} significant.

The periodograms of the Figs. 7.2, 7.3, 7.4, and 7.5 show peaks at various period lengths T.

For the yearly Hohenpeißenberg temperature means we notice larger periodogram values at longer periods (weakly significant at ≈ 15 years), see Fig. 7.2. Whether this is a true cycle or remainder of a trend, not successfully removed, is a matter of interpretation or may be clarified by further analyses. The non-significant period $T \approx 2.3$ had attained weak significance in Fig. 7.1, with regard to (the smoothed periodogram and) the AR(1)-corrected bounds. We have the problem, which periodogram version—without trend removal but with AR(1)-adjustment or with trend removal and without AR(1)-adjustment—should be preferred.

The periodogram of the Hohenpeißenberg winter series has maximum values, which are weakly significant, see Fig. 7.3. It is only the longest cycle of 15.3 years which is present in the yearly data, too.

For the Potsdam series of yearly data, we observe a significant peak at $T = 7.9$; the smoothed version confirms the interval $6 \leq T \leq 12$. This peak is also present in the periodogram of the Potsdam winter series. Here, the smoothed version distinguishes the same interval [6, 12], but at a lower level (Figs. 7.4, 7.5). The wavelet analysis will draw a more differentiated picture. The $T = 15.3$ cycle of the Hohenpeißenberg periodograms in Figs. 7.2 and 7.3 may be comprehended as

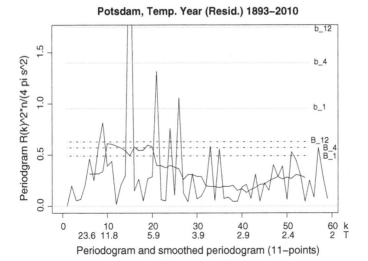

Fig. 7.4 Potsdam. Annual temperature means (detrended). Legend as in Fig. 7.2. The periodogram shows peaks ($>b_1$) at the periods of $T = 7.9, 5.6, 4.5$ (years) The maximum value is 3.06

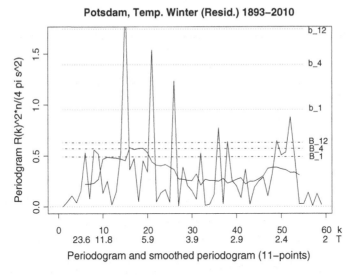

Fig. 7.5 Potsdam. Winter temperature means (detrended). Legend as in Fig. 7.2. The periodogram shows peaks ($>b_1$) at the periods of $T = 7.9, 5.6, 4.5$ (years). The maximum value is 2.23

the double cycle of Potsdam's $T = 7.9$. For Potsdam, but not for Hohenpeißenberg, the periodograms of the annual and of the winter temperature series exhibit great similarity.

In the periodograms of the temperature series, so far discussed, there is a side-peak at $T = 2.2 \dots 2.3$ years (except in Fig. 7.4). A possible meteorological explanation

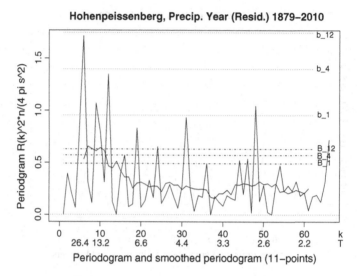

Fig. 7.6 Hohenpeißenberg. Annual precipitation (detrended). Legend as in Fig. 7.2. The periodogram shows peaks ($>b_1$) at the periods $T = 22, 14.7, 11, 2.75$ (years)

is the quasi-biannual (26 months) periodical oscillation (QBO) of the wind direction between east and west in the tropical stratosphere [see Schönwiese (1974), or Malberg (2007)]. A more trivial explanation is that of a possible artifact. The polynomial smoothing technique may leave behind a "zigzag" in the detrended series, creating a $T = 2$ cycle.

7.4 Precipitation. Summary

The periodogram of Hohenpeißenberg's annual precipitation series in Fig. 7.6 has a (nearly) significant peak at $T = 22$ years and a further peak at the half period length of $T = 11$ years. The smoothed version attains the B_{12} significance line between $T = 13$ and 20, which confirms the importance of the two peaks.

The periodogram of Fig. 7.7 for winter precipitation shows a narrow significant peak at $T \approx 4$ years. The smoothed version distributes it over the—not significant—interval [3.3, 4.3]; the wavelet analysis will shed more light on this point. Possibly, this period of 4 years is an (approximate) doubling of the number 2 (or 2.2), mentioned above. Note that the periodograms of the annual and of the winter precipitation series (for Hohenpeißenberg) exhibit no great resemblances. The same is true for Potsdam, as now follows, as well as for Bremen and Karlsruhe (no Figs., but see Table 7.1, right half).

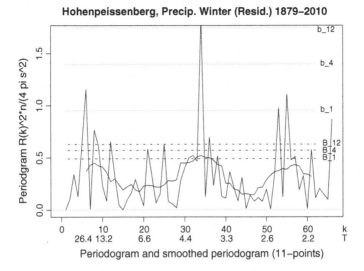

Fig. 7.7 Hohenpeißenberg. Winter precipitation (detrended). Legend as in Fig. 7.2. The periodogram shows peaks ($>b_1$) at the periods $T = 22, 3.9, 2.5, 2.4$ (years)

Table 7.1 Periodogram analysis of annual and winter temperature and precipitation series (detrended)

Station		Temperature	Precipitation
Bremen	Year	**8.1**, 7.6	4.7, 4.5, <u>4.2</u>
	Winter	<u>8.1</u>, 7.6, <u>5.8</u>, 2.3, 2.2	6.4, 4.8, 4.2, 3.3
Hohenpeißenberg	Year	17.7, <u>15.3</u>, 12.8, 7.9, 3.4, 2.3	<u>22</u>, 14.7, 11, 2.8
	Winter	<u>15.3</u>, <u>5.8</u>, 4.6, <u>2.6</u>, 2.3	22, **3.9**, 2.5, 2.4
Karlsruhe	Year	**105**, 8.4, <u>7.8</u>, 2.2	4.6, 4.2, 3.4
	Winter	<u>8.4</u>, 7.8, 5.5, 3.5, 3.1, 2.3	3.5, 3.2
Potsdam	Year	**7.9**, 5.6, 4.5	6.6, 4.2, 3.3, <u>2.3</u>, 2.1
	Winter	**7.9**, <u>5.6</u>, 4.5	9.8, <u>2.0</u>

Presented are the period lengths T (years) with a periodogram value exceeding the bound b_1. If the bound b_4 [b_{12}] is exceeded, the figure is underlined, e.g., <u>4.2</u> [put in boldface, e.g., **8.1**]

The periodogram of the Potsdam annual precipitation series shows a weak significant peak at $T = 2.3$ years (no Fig.). The plot in Fig. 7.8 of the winter precipitation has no significant values for $T > 2$. The largest value at the right margin (corresponding to $T = 2$) is perhaps a further hint at the meteorological phenomenon or at the artifact of the polynomial smoothing technique, both mentioned above at the end of 7.3

Summary. A summary of the preceding results is given in Table 7.1 (detrended series are considered only). The cycles with $T \approx 8$ years in the yearly and the winter temperature series Potsdam and Bremen are statistically significant; the same is true for the period $T \approx 4$ years in the winter precipitation series Hohenpeißenberg. The

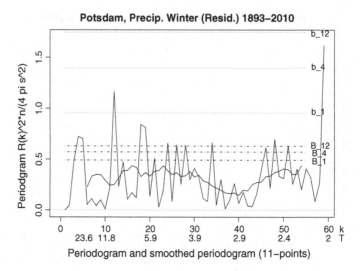

Potsdam, Precip. Winter (Resid.) 1893–2010

Periodogram and smoothed periodogram (11–points)

Fig. 7.8 Potsdam. Winter precipitation (detrended). Legend as in Fig. 7.2. The periodogram shows a peak ($>b_1$) at the period $T = 9.8$ (years) as well as a larger value at the right margin

Karlsruhe series contain one significant period, namely $T = 105$ years for annual temperatures (which one could also interpret as a long-time trend).

It is difficult to derive general statements (on periodicities in our climate series) from the Table 7.1 and the Figs. 7.2, 7.3, 7.4, 7.5, 7.6, 7.7, and 7.8. First of all, one has to mention the cycle of $T \approx 2.2$ years, discussed above. Then, in temperature series, there is a tendency for $T \approx 8$ and for $T = 5.5 \ldots 5.8$ years cycles. In precipitation series, $T \approx 4$ and $T \approx 3.3$ often appear in Table 7.1. These findings partly agree with those in Schönwiese (1974).

Once again, we want to point to some inherent problems of our analysis. Without trend removal, long-time periods may dominate (without belonging to true cycles); with trend removal, true cycles can be destroyed. Further, the assessment of significance in (smoothed) periodograms is not free from arbitrariness, nor is it the choice between the two versions, periodogram and smoothed periodogram.

R 7.1 Computation of the periodogram of the time series $Y[1], \ldots, Y[n]$ by calculating the Fourier coefficients $a[k]$ and $b[k]$, $k = 1, \ldots, n/2$. Plot of the (standardized) periodogram by means of the user function `plotP`, together with simultaneous bounds, see Fig. 7.7. Beneath the cycle numbers "k" we write the corresponding period lengths "T" by `mtext(..,line 2,..)`.

```
attach(hohenPr)
postscript(file="C:/CLIM/Hpgram.ps",height=6,width=15,horiz=F)
quot<-"Hohenpeissenberg, Precip. Winter (Resid.) 1879-2010";quot

#----------------------------------------------------------------
plotP<- function(k,z,nh,tylab,yli,bc,bct,lte,xte){
```

```
plot(k,z,type="l",lty=1,xlim=c(0,nh),ylim=yli,ylab=tylab,xlab="")
segments(1,bc,nh-4,bc,lty=2)              #Drawing simultaneous bounds
text(nh,bc,bct,cex=0.7)
#T-values as text on the bottom margin (side=1,line=2)
mtext(lte,side=1,line=2,at=xte[1:7])
mtext(c("k","T"),side=1,line=c(1,2),at=xte[8])
}
#----------Preparation of the vector Y, to be analyzed ---------
Y1<- (dcly+jan+feb)/1000    #Amount of precipitation winter [dm]
Ja<- Year-1900; Ja2<- Ja*Ja; Ja3<- Ja2*Ja; Ja4<- Ja2*Ja2
Y<- Y1 - predict(lm(Y1~Ja+Ja2+Ja3+Ja4))  #Removal of polyn.trend
n<- length(Y); nh<- round(n/2); SD2<- var(Y)

#Calculate periodogram via Fourier-coefficients a,b
seq<-1:nh; pg<-seq                         #vectors of dim nh
for (k in 1:nh)
{a<- 0; b<- 0; omk<- 2*pi*k/n
for (i in 1:n)
{a<- a+ Y[i]*cos(omk*i)
 b<- b+ Y[i]*sin(omk*i)
 pg[k]<- (a*a+b*b)/(n*pi)}
}

#-------------Plotting the periodogram------------------------
Pgr<- pg/SD2                               #Standardizing
tylab<- "Periodogram R(k)^2*n/(4 pi s^2)"
#Simultaneous bounds b_l=-ln(alpha/l)/pi, l=1,4,12
bc<- -log(0.05/c(1,4,12))/pi               #3 bounds b_1,b_4,b_12
bct<- c("b_1","b_4","b_12")
lte<- c("26.4","13.2","6.6","4.4","3.3","2.6","2.2")
xte<- c(5,10,20,30,40,50,60,70)
yli<- c(0.0,1.8)

plotP(seq,Pgr,nh,tylab,yli,bc,bct,lte,xte)
title(main=quot)

dev.off()
```

7.5 Wavelet Analysis

The cycles existing in a time series may have period lengths which vary in the course
of time. By means of wavelet analysis a spectrum can be established for each of the
ongoing time points. So we are able to trace the period lengths with maximal spectral
value along the time axis.

We choose a very specific wavelet method: Periodogram analysis is performed
within a Gaussian type window, which is moving from time point to time point
(*Morlet wavelets*). The wavelet spectrum is defined by

$$W(t, s) = A^2(t, s) + B^2(t, s), \quad t = 1, \ldots, n-1, \quad 0 < s < n. \tag{7.5}$$

With the abbreviations

$$a(\eta) = \cos(2\pi\eta), \quad b(\eta) = \sin(2\pi\eta), \quad f(\eta) = e^{-\eta^2/2},$$

the cos and sin terms in Eq. (7.5) are given by

$$A(t, s) = c_0 \cdot \frac{1}{\sqrt{s}} \cdot \sum_{t'=1}^{n} a(\eta(t', t, s)) \cdot f(\eta(t', t, s)) \cdot Y_{t'}$$

$$B(t, s) = c_0 \cdot \frac{1}{\sqrt{s}} \cdot \sum_{t'=1}^{n} b(\eta(t', t, s)) \cdot f(\eta(t', t, s)) \cdot Y_{t'}, \tag{7.6}$$

see Torrence and Compo (1998). Hereby we have set $c_0 = 1/\sqrt[4]{\pi}$ and

$$\eta(t', t, s) = \frac{t - t'}{s}.$$

Using the normalizing factor $c(s) = \frac{1}{\sqrt[4]{\pi}} \cdot \frac{1}{\sqrt{s}}$ from Eq. (7.6), we find that the function

$$g(t, s) = c(s) \cdot \exp\left(-\frac{1}{2}\left(\frac{t - t_0}{s}\right)^2\right) \quad \text{fulfills} \quad \int_{-\infty}^{\infty} g^2(t, s)\, dt = 1.$$

The wavelet spectrum $W(t, T_j)$ is evaluated at all time points $t = 1, \ldots, n-1$ and at certain periods $s = s_j = T_j$. With constants s_0 and δ, these periods are

$$s_j = T_j = s_0 \cdot 2^{j \cdot \delta}, \quad j = 0, \ldots, J-1.$$

We have the inverse relation

$$j = \frac{1}{\delta} \cdot \log_2\left(\frac{s_j}{s_0}\right).$$

Guided by the last equation, we put

$$J = \left[\frac{1}{\delta} \cdot \log_2\left(\frac{n}{s_0}\right)\right].$$

As constants we choose here $s_0 = 2$, $\delta = 0.5$. So we have the selected periods

$$T_0 = s_0 = 2, \quad T_1 = 2 \cdot \sqrt{2}, \quad T_2 = 4, \ldots, T_{J-1} = n/\sqrt{2},$$

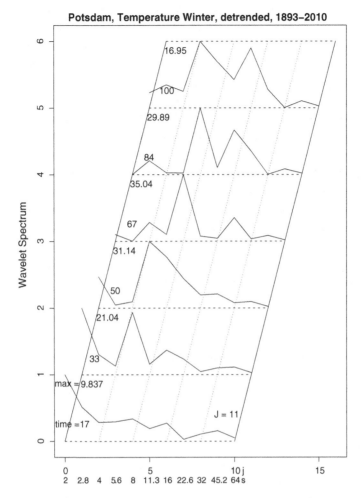

Fig. 7.9 Potsdam. Winter temperature. Wavelet spectrum W at six time points, equally spaced over the time interval $[0, n]$, $n = 116$, i.e., at $t = 17, 33, \ldots, 100$. The spectra are plotted in a normalized form $W / \max W \in [0, 1]$; the maximal spectral values $\max W$ are given for each of the six time points, that are $(9.837, \ldots, 16.95)$

the latter, if n is a power of 2. Wavelet spectra, each showing $W(t, T_j)$ for six time points t, equally spaced within the interval $[0, n]$, are presented in Figs. 7.9 and 7.11.

In the upper parts of Figs. 7.10 and 7.12, the index number $j(m) = j(t, m)$ is plotted over $t = 1, \ldots, n - 1$, where $T_{j(m)}$ is the period of maximal spectral value $W(t, T_j)$. This period $T = T_{j(m)}$ is called *dominant* period (periodicity) in the following. The lower plot presents the averaged spectrum, that is

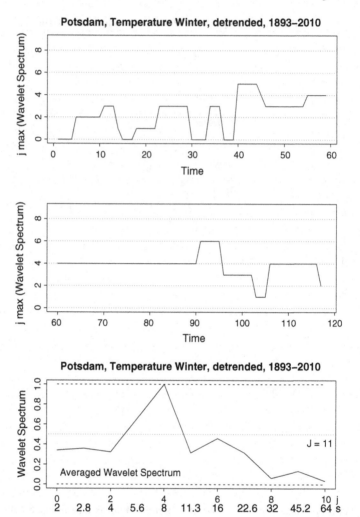

Fig. 7.10 Potsdam. Winter temperature. *Upper plots* Number $j(m)$, where the wavelet spectrum has the maximum value, for time points t, $0 \le t \le n - 1$. *Lower plot* Wavelet spectra are averaged over all time points t and plotted in the normalized form $W / \max W \in [0, 1]$. The maximal spectral value $\max W$ is 16.4

$$\frac{1}{n - 1} \cdot \sum_{t=1}^{n-1} W(t, T_j), \qquad j = 0, \dots, J - 1,$$

which can be compared with the (smoothed) periodograms of Sect. 7.1 (In our calculus we have neglected the constant factor c_0).

Detrended Winter Data

We will apply the wavelet method, presented above, to Potsdam's winter temperature
and to Hohenpeißenberg's winter precipitation. In both cases we are dealing with the
series after removal of a polynomial(4)-trend.

Potsdam, winter temperature: The diagrams of Figs. 7.5 and 7.10 (lower plot)
distinguish the period of $T = 8$ years (that is $j = 4$). However, it is present—
as dominant period—in the third quarter (and partly in the fourth quarter) of the
time interval $[0, n]$ only; at other times we have peaks at varying period lengths
(Fig. 7.9 and upper parts of Fig. 7.10). The wavelet analysis here reveals, in which
time intervals the cycle of $T = 8$ years is really present and in which not.

Hohenpeißenberg, winter precipitation: The periodogram of Fig. 7.7 has a maximal peak at period length $T = 4$ years, and a side-peak at $T = 22$. The wavelet
analysis reverses the roles, clearly expressed by Fig. 7.12 (lower plot). The period
of $T = 4$ (that is $j = 2$) is present—as dominant period—only at the beginning
and then at scattered succeeding time points (Fig. 7.11 and upper parts of Fig. 7.12);
mostly, the period length $T = 22.6$ ($j = 7$) is dominant.

R 7.2 Morlet wavelet spectrum $w[k, j]$, $k = 1, \ldots, n - 1$, $j = 1, \ldots, J$, of
the time series $Y[1], \ldots, Y[n]$, by means of the user function Wspectr (in the
following the constant factor c_0 is omitted). The vector Y is read from C:/CLIM/
HoPrWi.txt, containing the (detrended) winter precipitation amounts at Hohenpeißenberg. The $(n - 1) \times J$ matrix w is written on C:/CLIM/WaveOut.txt
and serves as input for further plotting and evaluation programs (by which Figs. 7.9,
7.10, 7.11, and 7.12 were produced).

```
Y<- scan("C:/CLIM/HoPrWi.txt")
n<- length(Y)

#---------Morlet Wavelet Spectrum---------------------------
Wspectr<- function(Y,n,s,J,t0,om){
spec<- 1:J                              #spec vector of dim J
for (j in 1:J){ sumc<- 0; sums<- 0
for (i in 1:n){ tij<- (t0 - i)/s[j]
sumc<- sumc + Y[i]*cos(om*tij)*exp(-0.5*(tij^2))
sums<- sums + Y[i]*sin(om*tij)*exp(-0.5*(tij^2)) }
spec[j]<- (sumc^2 + sums^2)/s[j]}
return(spec)
}


#-----------------------------------------------------------
om<- 2*pi; dj<- 0.50; s0<- 2              #Wavelet parameters
J<- trunc((1/dj)*log2(n/s0))
jot<- 0:(J-1); s<- s0*2^(jot*dj)          #jot,s vectors of dim J
c("om"=om,"dj"=dj,"s0"=s0,"J"=J)
"Vector s of length J"; s
```

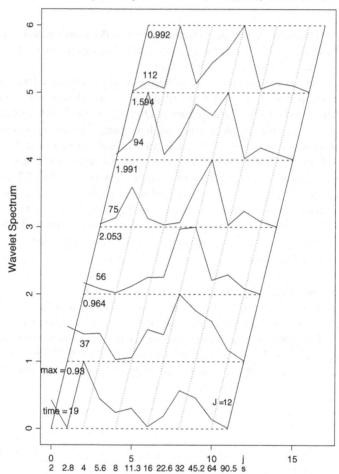

Fig. 7.11 Hohenpeißenberg. Winter precipitation. Wavelet spectrum W at six time points, equally spaced over the time interval $[0, n]$, $n = 130$, i.e. at $t = 19, 37, \ldots, 112$. The spectra are plotted in the normalized form $W / \max W \in [0, 1]$; the maximal spectral values $\max W$ are given for each of the six time points, that are $(0.93, \ldots, 0.992)$

```
sink("C:/CLIM/WaveOut.txt")
w<- 1:((n-1)*J); dim(w)<- c((n-1),J)    #w matrix of dim (n-1)xJ
for(k in 1:(n-1)){ spec<- Wspectr(Y,n,s,J,k,om)
w[k,]<- spec
write(w[k,],ncolumns=6,file="")
}
```

Output from R 7.2 Wavelet analysis for $n = 132$ winter precipitation amounts (detrended). The vector s consists of the selected $J = 12$ period lengths T. From the

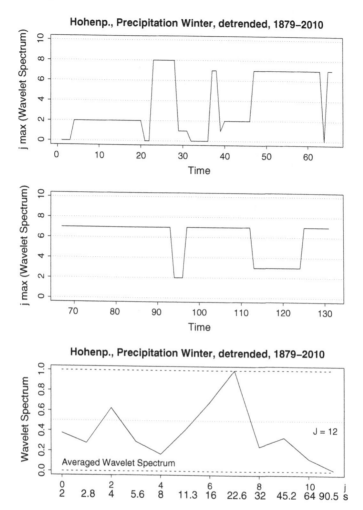

Fig. 7.12 Hohenpeißenberg. Winter precipitation. *Upper plots* Number $j(m)$, where the wavelet spectrum has the maximum value, for time points t, $0 \le t \le n-1$. *Lower plot* Wavelet spectra are averaged over all time points t and then plotted in the normalized form $W / \max W \in [0, 1]$. The maximal spectral value $\max W$ is 1.14

$(n-1) \times J$ matrix w, we reproduce here the first and the last 3 rows, each having 12 components.

```
om     dj    s0    J
6.283  0.50  2.00  12.00

"Vector s of length J"
2.000 2.828 4.000 5.657 8.000 11.314
16.000 22.627 32.000 45.255 64.000 90.510
```

WaveOut.txt
0.2636 0.0828 0.1084 0.0602 0.1301 0.1138
0.0185 0.0107 0.1998 0.2454 0.0921 0.0100
0.3567 0.1555 0.1715 0.0473 0.1381 0.1254
0.0191 0.0130 0.2141 0.2548 0.0943 0.0102
0.2931 0.2551 0.2540 0.0286 0.1415 0.1369
0.0196 0.0157 0.2290 0.2644 0.0966 0.0103
...   ...
0.0450 0.0648 0.0350 0.1857 0.0050 0.1588
0.2664 0.3468 0.0263 0.0592 0.0688 0.0095
0.0190 0.0637 0.0137 0.1409 0.0062 0.1409
0.2454 0.3188 0.0245 0.0558 0.0669 0.0093
0.0013 0.0525 0.0046 0.1043 0.0076 0.1238
0.2248 0.2924 0.0228 0.0526 0.0650 0.0092

Chapter 8
Complements

First we state that the separation of the trend/season component on one side and of the auto-correlation structure on the other side is crucial in our analysis. With regard to the latter: handling the detrended series or the differenced series by ARMA-type models was worked out in Chaps. 4 and 5. It should be mentioned that we have both aspects in mind, the modeling of the observed series and the predicting of climate values in the near future.

Alternatively to the ARMA-methods of Chaps. 4 and 5, we deal in this chapter with two approaches (growing polynomials, sin-/cos-approximation), which work without the separation mentioned above. With respect to annual data we introduce polynomials over growing time intervals, calculated for each interval anew. With respect to monthly data we approximate sin/cos functions, taking the sinusoidal form of the monthly temperature series for granted.

In addition, the 1-step predictions of Chap. 4 are extended to l-steps forecasts, $l = 2, 3, \ldots$ the number of years ahead.

We close with two special topics, the characterization of *temperature* versus *precipitation* variables and the relationship between *winter* and *yearly* data.

8.1 Annual Data: Growing Polynomials

According to the *forecast* approach, observations only up to time $t - 1$ are allowed for *predicting* a climate variable $Y(t)$ at time t. Accordingly, with regard to ARMA-equation (B.11), we did not proceed as usual, namely to estimate the α's and β's only once—for the whole sample. Rather, we proceeded step-by-step and estimated the coefficients for each time point t anew. Further, the moving averages were left-sided in the sense that only observations before time points t were involved. Polynomials, drawn only once, over the whole time interval $t = 1, \ldots, N$, however, were not qualified as an estimation and prediction method in Chaps. 4 and 5.

H. Pruscha, *Statistical Analysis of Climate Series*,
DOI: 10.1007/978-3-642-32084-2_8, © Springer-Verlag Berlin Heidelberg 2013

Table 8.1 Growing polynomial-prediction for annual temperature data, with RootMSQ-values for order numbers $m = 1, \ldots, 4$, and with the first three auto-correlation coefficients of the residual series in the case of $m = 2$

Station	RootMSQ					Auto-correlation of residuals		
	$m = 1$	$m = 2$	$m = 3$	$m = 4$	ARIMA	$r_e(1)$	$r_e(2)$	$r_e(3)$
Bremen	0.793	0.824	0.838	0.900	0.805	0.231	0.103	-0.181
Hohenpeißenberg	0.813	0.750	0.779	0.817	0.762	0.028	0.042	-0.069
Karlsruhe	0.792	0.705	0.723	0.759	0.687	0.155	0.105	-0.026
Potsdam	0.801	0.826	0.862	0.937	0.869	0.189	0.052	-0.228

For predicting annual climate variables, a possible alternative is the following method of polynomials over growing intervals (shortly: *growing* polynomials). For each $t = 1, \ldots, N$ anew, we fit the polynomial term

$$p_{t'}^{[t-1]} = a_0^{[t-1]} + a_1^{[t-1]} \cdot t' + \cdots + a_m^{[t-1]} \cdot (t')^m, \quad t' = 1, \ldots, t - 1,$$

of order m to the series $Y(t'), t' = 1, \ldots, t - 1$. The growing polynomial-prediction for $Y(t)$ is then given by

$$\hat{Y}(t) = p_t^{[t-1]}, \quad t = t_0 + 1, \ldots, N,$$

(once again, we choose $t_0 = [N/5]$ as the starting point for the predictions). Table 8.1 reports the RootMSQ-values for order numbers $m = 1, \ldots, 4$. Further, the first three auto-correlation coefficients $r_e(1), r_e(2), r_e(3)$ of the residual series

$$e(t) = Y(t) - \hat{Y}(t), \quad t = t_0 + 1, \ldots, N,$$

in the case of $m = 2$ are given. The coefficients $|r_e(h)|$ are (except for Hohenpeißenberg) in general larger than those of the ARMA-residuals of Table 4.4, and not sufficiently small in order to belong to a pure random series. According to the goodness-of-fit measure *RootMSQ*, the growing polynomials of order $m = 2$ perform better than those of order $m = 3$ or 4. A comparison of Figs. 8.2 and 8.3 reveals the reason: the $m = 4$ prediction follows closer the last year observation, i.e., it generally deviates more from the central course of the series (than the $m = 2$ prediction does), which turns out to be disadvantageous here. Figure 8.1 demonstrates this point in detail for polynomials over the years 1781–1988 (shown from 1960 onwards), with predictions for the year 1989.

The goodness-of-fit of this method is (for $m = 1$ or 2; except Karlsruhe) something better than that of the ARIMA-method, cf. Table 4.2. Its drawback: It gives no insight into the structure of the process. Further, when extending the method from 1-year to l-years predictions (as we do in the next section), it shows the disadvantage of polynomial extrapolation: the monotone and convex/concave divergence.

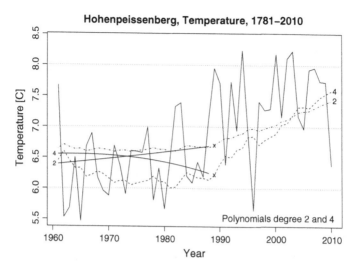

Fig. 8.1 Annual temperature means, observed (*solid zigzag line*), with predictions by *growing polynomials* of degrees 2 and 4 (*dashed-dotted line* and *dashed line*, resp., labels on the r.h.s.). Two polynomials (of orders 2 and 4) over the range 1781–1988 (*solid lines*, labels on the l.h.s.) are drawn, with predictions for the year 1989 (x). The last 50 years are shown

R 8.1 Fitting polynomials (order 4) over growing intervals [0, t], t = tst, ..., n (*growing* polynomials), and prediction for the time point t + 1 with the user function polypre.

```
attach(hohenTp)

#-------------------------------------------------------
polypre<- function(Y,n,x1,x2,x3,x4,tst){
polpr<- 1:(n+1)                          #vector of dim n+1
polpr[1:tst]<- mean(Y[1:tst])
for (t in tst:n)
{poly<-lm(Y[1:t]~x1[1:t]+x2[1:t]+x3[1:t]+x4[1:t])
b<-poly$coefficients                     #$b vector of dim 5
t1<-t+1
polpr[t1]<-b[1]+b[2]*x1[t1]+b[3]*x2[t1]+b[4]*x3[t1]+b[5]*x4[t1]
}
return(polpr)
}


#----Preparing the input of function polypre--------------------
n<- length(Year); tst<- trunc(n/5); ts1<-tst+1
Y<-Tyear/100
x1<-1:(n+1)-n/2; x2<-x1*x1;  x3<-x2*x1; x4<-x2*x2

Ypred<- polypre(Y,n,x1,x2,x3,x4,tst)
```

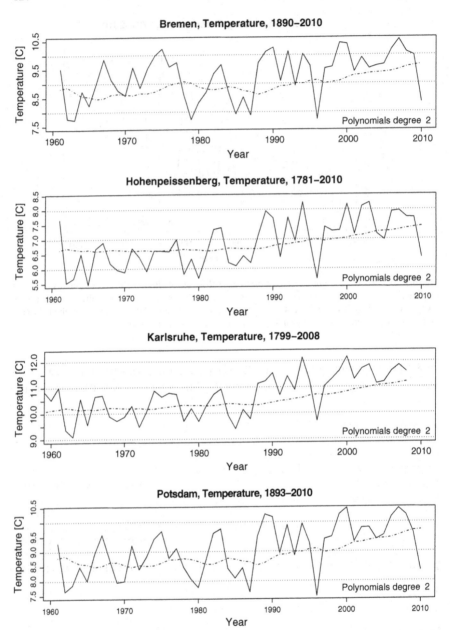

Fig. 8.2 Annual temperature means, observed (*solid line*) and predicted by *growing polynomials* of degree 2 (*dashed-dotted line*). The last 50 years are shown

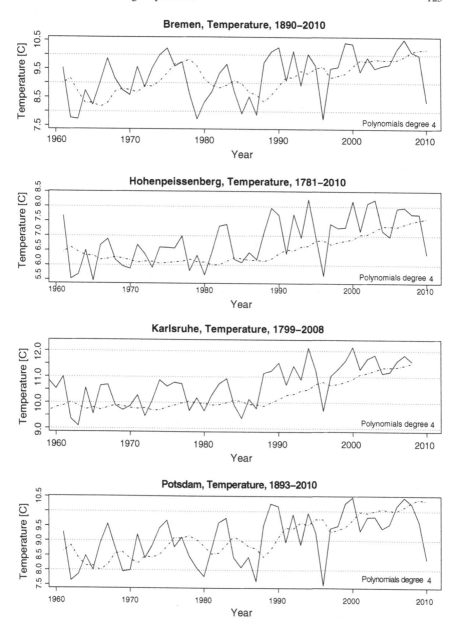

Fig. 8.3 Annual temperature means, observed (*solid line*) and predicted by *growing polynomials* of degree 4 (*dashed-dotted line*). The last 50 years are shown

```
"Poly_t-prediction last decade";  Ypred[(n-9):n]
Ypre<-Ypred[ts1:n]; Yres<-Y[ts1:n]-Ypre; MSQ<-mean(Yres*Yres)
c("std Res"=sqrt(var(Yres)), "MSQ"=MSQ,"rootMSQ"=sqrt(MSQ))
```

8.2 Annual Data: ARIMA l-Years Forecast

The ARIMA-prediction for the year 2011, on the basis of observations up to year
2010 (as done in Chap. 4), will now be called 1-step forecast. Box and Jenkins (1976)
gave forecast formulas for l-steps (here: for l years) ahead, when an ARMA-model
is assumed; see Eqs. (B.18)–(B.20).
    Let us denote by

$$X(t) = Y(t) - Y(t - 1), \quad t = 2, \ldots, N, \quad [X(1) = 0]$$

as in Sect. 4.1 the differenced series, i.e., the series of the annual changes in the
temperature mean. An ARMA(p, q) model is fitted to $X(t)$, yielding coefficients
$\alpha_i = \alpha_i^{[1,N]}$ and $\beta_j = \beta_j^{[1,N]}$. Using these coefficients, forecasts $\hat{X}_N(1), \ldots, \hat{X}_N(l)$
for the variables $X(N + 1), \ldots, X(N + l)$ are calculated (put $T = N$ and $Y = X$
in (B.18)–(B.20)). Then the ARIMA-forecasts for the integrated process $Y(t)$ are
iteratively gained by

$$\hat{Y}_N(1) = Y(N) + \hat{X}_N(1), \ldots, \hat{Y}_N(l) = \hat{Y}_N(l - 1) + \hat{X}_N(l). \tag{8.1}$$

The results, gained below with Monte Carlo simulations, will justify this approach
to a certain extent. Figures 8.4 and 8.5 present the ARIMA-forecasts (8.1), together
with lower and upper interval boundaries for $\alpha = 0.2, 0.4$, gained by the Monte
Carlo method. It should be noted that these forecast intervals are intervals for a
single random variable and are not to be mixed up with confidence intervals: from
there the relatively large α values and the relatively large intervals around the $\hat{Y}_N$'s.
    Instead of using the Box–Jenkins forecast function, we now apply the Monte
Carlo method to gain forecasts, together with $(1 - \alpha)$-probability intervals. For this
purpose, one uses the recursive ARIMA-equations

$$\begin{aligned} X(N + k) &= \alpha_p X(N + k - p) + \cdots + \alpha_1 X(N + k - 1) \\ &\quad + \beta_q e(N + k - q) + \cdots + \beta_1 e(N + k - 1) + e(N + k), \\ Y(N + k) &= Y(N + k - 1) + X(N + k), \quad k = 1, \ldots, l. \end{aligned} \tag{8.2}$$

Hereby, the error terms $e(t)$, $t \leq N$, have to be iteratively calculated, and $e(N + 1), \ldots, e(N + l)$ are drawn as $N(0, s_e^2)$-distributed random numbers; s_e^2 being the
variance of the $e(t)$, $t \leq N$. Further, variables $X(t)$ up to time point N are observa-
tions (here: differences thereof), variables $X(t)$ after N are calculated by Eq. (8.2).
This yields us one single Monte Carlo simulation of an ARIMA-forecast. Doing

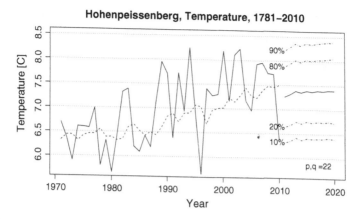

Fig. 8.4 Hohenpeißenberg, temperature 1781–2010. Time series of annual means (°C) (*solid zigzag line*), together with the ARIMA-predictions up to year 2010 (*dashed line*; the last 40 years are shown). Subsequently, ARIMA-forecasts (8.1) acc. to Box and Jenkins for the next 10 years 2011 till 2020 (*solid line*), together with Monte Carlo forecast boundaries

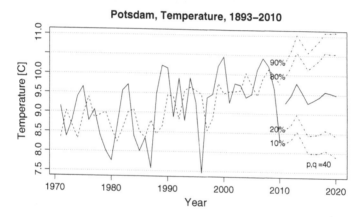

Fig. 8.5 Potsdam, temperature 1893–2010. Time series of annual means. Same legend as in Fig. 8.4

this MC times (we worked with $MC = 20000$), we build—for each $k = 1, \ldots, l$ separately—the mean value $\tilde{Y}(N + k)$ and quantiles

$$\tilde{Q}_\beta(N + k), \quad \beta = 0.10, 0.20, 0.80, 0.90.$$

We consider the Monte Carlo quantile curves as being close to the "true" quantile curves.

Figures 8.6 and 8.7 tell us that the Box–Jenkins function $\hat{Y}_N(k)$, $k = 1, \ldots, l$, gained from one single application, is nearly identical with the Monte Carlo forecast

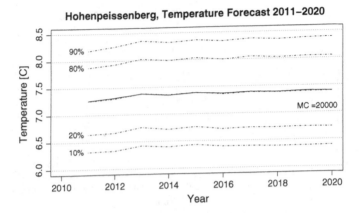

Fig. 8.6 Hohenpeißenberg, temperature 1781–2010. The ARIMA-forecasts for the next 10 years 2011 till 2020, together with forecast boundaries: the mean and the quantiles of 20000 Monte-Carlo repetitions (*dashed-dotted lines*), the Box–Jenkins forecast function (8.1) (*solid line*)

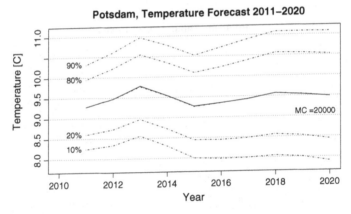

Fig. 8.7 Potsdam, temperature 1893–2010. The ARIMA-forecasts for the next 10 years 2011 till 2020. Same legend as in Fig. 8.6

function $\tilde{Y}(N + k)$ (gained from many replications of Eq. (8.2)). This is due to the linearity of Eq. (8.2) and of the conditional expectation (B.15). As one expects on account of the second part of Eq. (8.2), the "true" interval bounds $\tilde{Q}_\beta(N + k)$, $k = 1, \ldots, l$, slowly diverge.

Remark. The Monte Carlo forecast method will be especially valuable, when a nonlinear equation governs the evolution of the process.

R 8.2 Box–Jenkins forecast method, by using the two user functions `epsilon`, `forec`. The time series vector `Y` is read from `C:/CLIM/TimeS.txt`, here the yearly temperature series of Hohenpeißenberg. It is transformed into the series `X` by differencing. (If the vector `Y` needs no differencing, then the lines ending with

♯ − − ♯ should be omitted, and we have X = Y, Xfore = Yfore). In forec, new error terms are entering as zeros, in monca—see **R 8.3** below—they are drawn as random numbers.

```
Y<- scan("C:/CLIM/TimeS.txt")
"Hohenpeissenberg, Temperature, 1781-2010"                    #--#
library(TSA)                              #see CRAN software-packages

#-------------------------------------------------------------
epsilon<- function(y,n,mc,theta,a,b){  #error terms recursively
ypr<- rep(mean(y),times=n); eps<- rep(0,times=n)  #ve. of dim n
for (t in (mc+1):n){
Suma<- theta; Sumb<- 0
for (m in 1:mc){
Suma<- Suma+a[m]*y[t-m]; Sumb<- Sumb+b[m]*eps[t-m]}
ypr[t]<- Suma + Sumb
eps[t]<- y[t] - ypr[t]}
return(eps)
}


forec<- function(y,e,n,mc,theta,a,b,la){ #BJ forecast function
fore<- rep(mean(y),times=la)                  #vector of dim la
ep<- e, yp<- y                                #vectors of dim n
for(j in 1:la){fore[j]<- theta
for(k in 1:mc){fore[j]<- fore[j]+a[k]*yp[n-k+1]+b[k]*ep[n-k+1]}
yp[1:(n-1)]<- yp[2:n]; yp[n]<- fore[j]
ep[1:(n-1)]<- ep[2:n]; ep[n]<- 0              #new error term = 0
}
return(fore)
}


#------Data----------------------------------------------------
N<-length(Y);  X<- Y                          #Y = time series
X[1]<-0; X[2:N]<-Y[2:N]-Y[1:(N-1)]  #X=differenced series #--#
ma<- 2; mb<-2; mc<- max(ma,mb)                #mc maximal 6
c("ArOrder"=ma,"MAOrder"=mb)

# --------- ARMA(p,q)-Model for series X --------------------
xarma<- arma(X,order=c(ma,mb))
summary(xarma)

xcoef<- xarma$coef                    #$vector of dim ma+mb+1
#Coefficients a,b,theta
a<- rep(0,times=6); b<- rep(0,times=6)
if (ma > 0) {for (m in 1:ma){a[m]<- xcoef[m]}}
if (mb > 0) {for (m in 1:mb){b[m]<- xcoef[ma+m]}}
theta<- xcoef[ma+mb+1]

epsil<- epsilon(X,N,mc,theta,a,b) #user function: error terms

#--------Forecasting acc. to Box&Jenkins----------------------
la<-10
```

```
#user function forec: ARMA forecast function Xfore
Xfore<-forec(X,epsil,N,mc,theta,a,b,la)
"Box&Jenkins ARMA forecast vector (X series)"
Xfore; Yfore<- Xfore

#ARIMA forecast function Yfore (integrated series)         #--#
Yfore<- Y[N] + Xfore                                       #--#
if(la > 1) for(j in 2:la){Yfore[j]<-Yfore[j-1]+Xfore[j]}   #--#
"Box&Jenkins ARIMA forecast vector (Y series)"; Yfore      #--#
```

Output from R 8.2 The coefficients $\alpha_i = \alpha_i^{[1,N]}$, $\beta_j = \beta_j^{[1,N]}$, given below, are computed for the whole series, see Table 4.2. Recall that X denotes the differenced, Y the integrated series. A plot of the forecast vector can be found in Fig. 8.4.

```
"Hohenpeissenberg, Temperature, 1781-2010"
ArOrder MAOrder
    2        2
Model:  ARMA(2,2)


Coefficient(s):
            Estimate   Std. Error   t value  Pr(>|t|)
ar1         -0.63915      0.14671     -4.36   1.3e-05  ***
ar2          0.10524      0.06780      1.55     0.12
ma1         -0.17896      0.14223     -1.26     0.21
ma2         -0.67221      0.13493     -4.98   6.3e-07  ***
intercept    0.00834      0.00788      1.06     0.29


  "Box&Jenkins ARMA forecast vector (X series)"
  0.8886 0.050 0.0698 -0.031 0.0355 -0.0176 0.0233 -0.0084 0.016
-0.0029
  "Box&Jenkins ARIMA forecast vector (Y series)"
  7.2636 7.3138 7.3836 7.3526 7.3881 7.3705 7.3938 7.3854 7.4016
  7.3987
```

R 8.3 Monte-Carlo forecast method, by using the two user functions epsilon, monca. The time series vector Y is read from C:/CLIM/TimeS.txt, here the yearly temperature series of Hohenpeißenberg. It is transformed into the series X by differencing. (If the vector Y needs no differencing, then the lines ending with ♯ − −♯ should be omitted, and we have X = Y, Xmonca = Ymonca). In the program **R 8.2** above, function forec, new error terms are entered as zeros, in monca they are drawn as random numbers rnorm.

```
Y<- scan("C:/CLIM/TimeS.txt")
"Hohenpeissenberg, Temperature,   1781-2010"                    #--#
library(TSA)                               #see CRAN software-packages

#------------------------------------------------------------------
epsilon<- function(y,n,mc,theta,a,b){  #error terms recursively
ypr<- rep(mean(y),times=n); eps<- rep(0,times=n)  #ve. of dim n
```

```
for   (t in (mc+1):n){
Suma<- theta; Sumb<- 0
for (m in 1:mc){
Suma<- Suma+a[m]*y[t-m]; Sumb<- Sumb+b[m]*eps[t-m]}
ypr[t]<- Suma + Sumb
eps[t]<- y[t] - ypr[t]}
return(eps)
}

monca<- function(y,e,n,mc,theta,a,b,la,se){        #MC simulation
monc<- rep(mean(y),times=la)
ep<- e ; yp<- y; eps<- rnorm(la)         #la N(0,1) random numbers
for(j in 1:la){monc[j]<- theta+eps[j]*se
for(k in 1:mc){monc[j]<- monc[j]+a[k]*yp[n-k+1]+b[k]*ep[n-k+1]}
yp[1:(n-1)]<- yp[2:n]; yp[n]<- monc[j]
ep[1:(n-1)]<- ep[2:n]; ep[n]<- eps[j]*se             #new error term
}
return(monc)
}

#------Data----------------------------------------------------
N<-length(Y);   X<- Y                          #Y = time series
X[1]<-0; X[2:N]<-Y[2:N]-Y[1:(N-1)]   #X=differenced series #--#
ma<- 2; mb<-2; mc<- max(ma,mb)                   #mc maximal 6
c("ArOrder"=ma,"MAOrder"=mb)

# ---------- ARMA(p,q)-Model for series X --------------------
xarma<- arma(X,order=c(ma,mb))
summary(xarma)

xcoef<- xarma$coef                     #$vector of dim ma+mb+1
#Coefficients a,b,theta
a<- rep(0,times=6); b<- rep(0,times=6)
if (ma > 0) {for (m in 1:ma){a[m]<- xcoef[m]}}
if (mb > 0) {for (m in 1:mb){b[m]<- xcoef[ma+m]}}
theta<- xcoef[ma+mb+1]

epsil<- epsilon(X,N,mc,theta,a,b) #user function: error terms
se<- sqrt(var(epsil[(mc+1):N]))    #error terms from mc+1 onw.
c("Mean Epsilon"=mean(epsil[(mc+1):N]),  "StdDev Epsilon"=se)

#--------Forecasting acc. to Monte Carlo----------------------
la<- 10; MC<- 20000; c("Monte Carlo Repetitions"=MC)
Xmonca<-1:(MC*la); dim(Xmonca)<- c(MC,la)  #MCxla matrix Xmonca
#user function monca: 1 Monte-Carlo repetition
for (m in 1:MC){
montc<- monca(X,epsil,N,mc,theta,a,b,la,se)
Xmonca[m,]<- montc }
Ymonca<- Xmonca

#Monte Carlo simulations Ymonca (integrated series)       #--#
Ymonca<- Y[N] + Xmonca                                    #--#
```

```
if(la>1) for(j in (2:la))                                        #--#
            {Ymonca[,j]<- Ymonca[,(j-1)]+Xmonca[,j]}     #--#
```

```
meYmonca<- colMeans(Ymonca)
"Monte Carlo mean vector (Y series)";   meYmonca
quan<- 1:(4*la); dim(quan)<- c(4,la)              #4 x la matrix quan
alph<- c(0.10,0.20,0.80,0.90)
for(j in 1:la){
quan[,j] <- quantile(Ymonca[,j],alph) }        #empirical quantiles
```

```
"Monte Carlo quantile vectors, levels 0.10, 0.20, 0.80, 0.90"
quan[,]
```

Output from R 8.3 The coefficients $\alpha_i = \alpha_i^{[1,N]}$, $\beta_j = \beta_j^{[1,N]}$, given below, are computed for the whole series, see Table 4.2. A plot of the following vectors can be found in Fig. 8.6.

```
"Hohenpeissenberg, Temperature, 1781-2010"
ArOrder MAOrder
      2       2
Model:  ARMA(2,2)
```

```
Coefficient(s):
              Estimate  Std. Error  t value Pr(>|t|)
ar1          -0.63915     0.14671    -4.36  1.3e-05 ***
ar2           0.10524     0.06780     1.55     0.12
ma1          -0.17896     0.14223    -1.26     0.21
ma2          -0.67221     0.13493    -4.98  6.3e-07 ***
intercept     0.00834     0.00788     1.06     0.29
```

```
Mean Epsilon    StdDev Epsilon
   0.0059          0.7700
```

```
Monte Carlo Repetitions    20000
 "Monte Carlo mean vector (Y series)"
 7.2729 7.3195 7.3814 7.3539 7.3930 7.3766 7.4043 7.3851 7.4008
 7.3977
```

```
 "Monte Carlo quantile vectors, levels 0.10, 0.20, 0.80, 0.90"
 6.341 6.358 6.425 6.402 6.414 6.417 6.409 6.397 6.419 6.412
 6.659 6.683 6.759 6.717 6.752 6.744 6.760 6.738 6.750 6.756
 7.890 7.949 8.010 7.979 8.026 8.018 8.059 8.039 8.054 8.045
 8.212 8.268 8.340 8.310 8.377 8.359 8.386 8.370 8.389 8.378
```

Table 8.2 Sin-/cos-modeling for monthly temperature data

Station	$a^{[1,M]}$	$b^{[1,M]}$	RootMSQ	$r_e(1)$	$r_e(2)$	$r_e(3)$
Bremen	−4.719	−6.921	1.930 (1.906)	0.293	0.132	0.082
Hohenpeißenberg	−5.029	−7.109	2.136 (2.141)	0.130	0.048	−0.010
Karlsruhe	−4.705	−8.050	1.908 (1.909)	0.172	0.061	−0.010
Potsdam	−5.033	−8.084	2.076 (2.013)	0.293	0.159	0.096

Coefficients, goodness-of-fit measure RootMSQ (in parenthesis the values for ARMA) and the first three auto-correlation coefficients of the residual series

Table 8.3 Sin-/cos-modeling for monthly precipitation data

Station	$a^{[1,M]}$	$b^{[1,M]}$	RootMSQ	$r_e(1)$	$r_e(2)$	$r_e(3)$
Bremen	−0.993	−0.686	2.995 (3.104)	0.039	0.013	0.008
Hohenpeißenberg	−2.345	−4.741	4.644 (4.710)	0.064	−0.006	−0.005
Karlsruhe	−0.707	−0.823	3.569 (3.621)	0.067	0.039	0.005
Potsdam	−0.704	−0.907	2.850 (2.945)	0.058	−0.007	−0.028

Caption as in Table 8.2

8.3 Monthly Data: Sin-/Cos-Modeling

For predicting *monthly* climate variables one has to tune the estimation of the trend and of the seasonal component. The succession of the ARIMA-method for the yearly trend and of the ARMA-method for the detrended series (as done in Chap. 5) leaves behind residuals which are close to a random series; see Table 5.3 above. This is not the case with the following method (which, however, shows a very good fit).

Like the ARIMA-trend+ARMA method of Chap. 5, the present method uses the yearly trend estimation by ARIMA. As an alternative to ARMA we now apply sin and cos functions for modeling and predicting. For each time point (month) t anew, $t = 1, \ldots, M = 12 * N$, we fit the harmonic term

$$s_{t'}^{[t-1]} = a^{[1,t-1]} \cdot \sin\left(\omega \cdot t'\right) + b^{[1,t-1]} \cdot \cos\left(\omega \cdot t'\right), \quad t' = 1, \ldots, t-1,$$

to the detrended series $X(t')$, $t' = 1, \ldots, t-1$, where $\omega = (2 \cdot \pi)/12$. The sin-/cos-prediction for $X(t)$ is then given by

$$\hat{X}(t) = s_t^{[t-1]}, \quad t = t_0 + 1, \ldots, M,$$

(analogously to Chap. 5, we choose $t_0 = [N/5] * 12$ as starting point for the predictions). Tables 8.2 and 8.3 report the coefficients $a^{[1,M]}$, $b^{[1,M]}$, calculated from the whole series, the *RootMSQ*-values and the first three auto-correlation coefficients $r_e(1), r_e(2), r_e(3)$ of the residual series

$$e(t) = X(t) - \hat{X}(t), \quad t = t_0 + 1, \ldots, M.$$

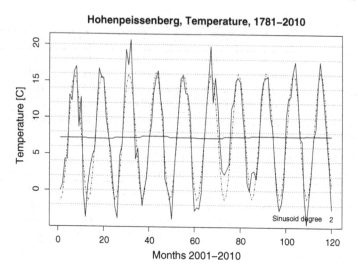

Fig. 8.8 Hohenpeißenberg, 1781–2010. Monthly temperature means (°C) (*solid zigzag line*), together with trend (*inner solid line*) and trend+sin-/cos-prediction (*dashed-dotted line*). The last 10 years are shown

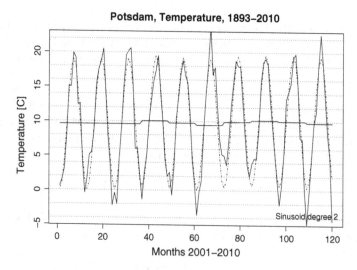

Fig. 8.9 Potsdam, 1893–2010. Same legend as in Fig. 8.8

The *RootMSQ*-values of the Tables 5.1 and 5.4 are added in parenthesis. For temperature, the first coefficient $r_e(1)$ is rather large and significantly different from zero, which rejects the assumption of a pure random series $e(t)$.

The goodness of fit of this sin-/cos-prediction is very satisfactory and quite close to (and mostly slightly better than) that of the ARMA-method (see also Figs. 8.8, 8.9, 8.10 and 8.11 for time series plots). It involves only two unknown

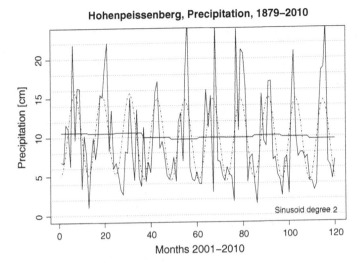

Fig. 8.10 Hohenpeißenberg, 1879–2010. Monthly precipitation amounts (cm) (*solid zigzag line*), together with trend (*inner solid line*) and trend+sin-/cos-prediction (*dashed-dotted line*). The last 10 years are shown

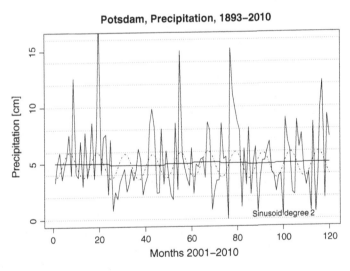

Fig. 8.11 Potsdam, 1893–2010. Same legend as in Fig. 8.10

coefficients—instead of four or five (as ARMA does in Chap. 5)—for fitting the detrended series. But one has to take into account, that the sin-/cos-model uses information about the sinusoidal form of the seasonal component of temperature and precipitation, and may be less suitable for other climate variables. ARMA-modeling is, much more than the sin-/cos-approach, a universal method.

Table 8.4 Correlation analysis for annual climate data on temperature (Temp.) and precipitation (Prec.), with lagged variables

Station	$r(Y, Y1)$		$r(Y, (Y1\ldots Y6))$		$r(Y, (Y1\ldots Y6, Z1\ldots Z6))$	
	Y = Temp.	Y = Prec.	Y = Temp.	Y = Prec.	Y = Temp. Z = Prec.	Y = Prec. Z = Temp.
Bremen	0.349	0.052	0.350	0.258	0.375	0.283
Hohenpeißenberg	0.296	0.274	0.375	0.319	0.539	0.359
Karlsruhe	0.332	0.009	0.455	0.136	0.665	0.238
Potsdam	0.356	−0.079	0.401	0.262	0.427	0.287

Coefficients of correlation and of multiple correlation are presented

A comparison of Figs. 8.8, 8.9 with Figs. 8.10, 8.11 shows that the model fits much better in the case of temperature than in the case of precipitation. The reason is the seasonal component, which is more distinct for the first than for the second climate variable.

This topic, i.e., *temperature* versus *precipitation* data, has already been discussed in Sects. 3.1 and 5.4, and will be further elaborated in the next section.

8.4 Further Topics

Temperature ↔ Precipitation

Our correlation and prediction analysis has revealed, that

(PT) *precipitation* is more irregular and closer to a pure random
       phenomenon than *temperature* is;

see also von Storch and Navarra (1993). This statement is also confirmed by Table 8.4, where the coefficients of correlation $r(Y, Y1)$ and of multiple correlation $r(Y, (Y1\ldots Y6))$ are presented. In this table, Y stands for annual temperature or for annual precipitation, and by $Y1$–$Y6$ we denote lagged variables, from lag = 1 to 6 years. If Y is, for example, the annual temperature mean, then $Y1, Y2, \ldots, Y6$ are the annual temperature means one, two, ..., six years before.

In Bremen, Karlsruhe, and Potsdam, the correlations between temperature variables are larger than those between precipitation variables; in Karlsruhe and Potsdam they are even distinctly larger. If precipitation (Y = Prec.) is correlated with the set $(Y1\ldots Y6, Z1\ldots Z6)$, comprising the lagged precipitation variables $Y1\ldots Y6$ *and* the lagged temperature variables $Z1\ldots Z6$, the coefficient remains—nevertheless— (far) below that of temperature (Y = Temp.), when correlated with the lagged temperature variables; compare the last column with the third (numerical) column.

The exception is Hohenpeißenberg, the station in the foreland of the Alps. Here, in comparison with the other three stations, the level of correlation for precipitation (Y = Prec.) is larger and—with regard to $r(Y, Y1)$ and $r(Y, (Y1\ldots Y6))$—closer to that for temperature (Y = Temp.).

Table 8.5 Correlation analysis for monthly climate data, seasonally adjusted, on temperature (Temp.) and precipitation (Prec.), with lagged variables

Station	$r(Y, Y1)$		$r(Y, (Y1 \ldots Y6))$		$r(Y, (Y1 \ldots Y6, Z1 \ldots Z6))$	
	$Y = $ Temp.	$Y = $ Prec.	$Y = $ Temp.	$Y = $ Prec.	$Y = $ Temp. $Z = $ Prec.	$Y = $ Prec. $Z = $ Temp.
Bremen	0.272	0.027	0.283	0.043	0.287	0.074
Hohenpeißenberg	0.153	0.013	0.177	0.057	0.200	0.106
Karlsruhe	0.197	0.029	0.223	0.047	0.252	0.066
Potsdam	0.276	0.005	0.288	0.050	0.298	0.061

Coefficients of correlation and of multiple correlation are presented

Analogously with Table 8.4, the Table 8.5 shows the (multiple) correlation coefficients for (seasonally adjusted) *monthly* climate data. Once again, they are much smaller for precipitation than for temperature. A special role of Hohenpeißenberg's precipitation data can be detected (at most) in the last column.

Results on the standardized RootMSQ value rsq—obtained in Chap. 5 for the trend + ARMA approach—substantiate the statement (PT) above. We have found in (5.3) and (5.4)

Monthly data	rsq-values			
	Bremen	Hohenp.	Karlsruhe	Potsdam
Temperature	0.305	0.330	0.280	0.283
Precipitation	0.999	0.791	0.996	0.999

The rsq coefficient as a measure of goodness of fit is much better for monthly temperature than for monthly precipitation; and for the latter, it is better in the case of Hohenpeißenberg than in the case of the other three stations. By analogy with standard regression analysis we can write

$$ \mathrm{rsq}^2 = 1 - R^2, $$

where R^2 is called *coefficient of determination* (and where R turns out to be a coefficient of multiple correlation). The value $R^2 \approx 0$ for the monthly precipitation data of Bremen, Karlsruhe, and Potsdam signalizes nearly total indetermination in these series.

For *annual* series these R^2-values are more balanced than for monthly data (see Sects. 4.2 and 4.4) with values roughly in the interval $0.2 \ldots 0.4$, for both, temperature *and* precipitation.

Table 8.6 Correlation coefficient r between climate variables, referring to winter and remaining year (i.e., here, the year from March to November), together with the level 0.05-bound b_1

	Hohenpeißenberg $n(Tp) = 230$ $n(Pr) = 132$			Karlsruhe $n(Tp) = 210$ $n(Pr) = 133$			Potsdam $n(Tp) = 118$ $n(Pr) = 118$		
	r	r	b_1	r	r	b_1	r	r	b_1
TpWi $\to$ TpYe⁻	(0.161)	0.068	0.13	(0.175)	0.107	0.14	(0.194)	0.167	0.18
PrWi $\to$ PrYe⁻	(0.281)	0.228	0.17	(0.059)	0.072	0.17	(0.116)	0.145	0.18
TpWi $\to$ PrYe⁻	(0.195)	0.150	0.17	(0.080)	0.089	0.17	(0.091)	0.113	0.18
PrWi $\to$ TpYe⁻	(0.033)	−0.07	0.17	(0.050)	−0.08	0.17	(0.068)	0.000	0.18

Variables are winter temperature (TpWi), winter precipitation (PrWi), temperature (TpYe⁻), and precipitation (PrYe⁻) of the remaining year; without trend removal (in parenthesis) and with trend removal

Table 8.7 Cross-correlation function for two pairs of variables, referring to (winter, remaining Year), after trend removal

Time lag	0	1	2	3	4	5	6	7	8
TpWi $\to$ TpYe⁻	0.068	0.075	0.077	−0.026	−0.022	−0.042	−0.034	−0.07	−0.01
PrWi $\to$ PrYe⁻	0.228	0.062	0.057	0.050	0.083	−0.079	−0.119	−0.16	0.02

The pairs are (TpWi,TpYe⁻) and (PrWi,PrYe⁻), the time lags are $0, 1, \ldots, 8$ years. Hohenpeißenberg. The simultaneous bounds are $b_8 = 0.180$ and 0.238, resp. See also the caption of Table 8.6

Winter $\leftrightarrow$ (Remaining) Year

Winter data are often considered as an indicator of the general climate development. In the following, we consider climatic variables, referring to

> *winter*, that are the months of December last year, January, February, and
> *remaining year*, that are March to November.

Looking in Table 8.6 for significant correlations (*after* trend removal), we find one single case only, namely in the Hohenpeißenberg series

> precipitation winter $\to$ precipitation remaining year ($r = 0.228$).

Next, we extend the correlation coefficients of Table 8.6 to cross-correlation functions, measuring over time lags of $1, 2, \ldots, 8$ years. We learn from Table 8.7, that then the significance—named above—disappears.

Looking back on Chaps. 2 and 7

The spectral analysis methods have been applied both to winter data and to annual data (now covering the whole year). A real correspondence between them has been discovered only in one case: the Potsdam annual and winter temperature series have the same significant cycle of $T = 7.8$ years, cf. Figs. 7.4 and 7.5. Additionally, in the Hohenpeißenberg winter and annual temperature series, that is in Figs. 7.2 and 7.3, we have the same weak significant cycle of $T = 15.3$ years (possibly a doubling of the Potsdam cycle). See also Pruscha (1986) for more spectral functions of winter and of annual climate data.

The development of the winter temperature in the last two centuries is less distinct than that of the yearly temperature. The warming in the winter months of the last decades is present, but it is modest compared with the corresponding yearly warming (see end of Sect. 2.2).

So we have to state that winter data alone are a weak indicator of the climate in the whole year and also—presumably—of the general climate development.

Appendix A
Excerpt from Climate Data Sets

We present excerpts from data sets, used in the preceding text. First the monthly temperature means of the years 1781–2010 and precipitation amounts of the years 1879–2010 at the station Hohenpeißenberg are given. These files are used in the program R 1.1 under the names HohenT.txt and HohenP.txt, resp. Then for five stations each, the annual temperature means and precipitation amounts (of the years 1930–2008) and the daily temperature and precipitation records (of the years 2004–2010) are reproduced. These files are named Years5.txt and Days5.txt in the R programs of Chaps. 3 and 6, respectively.

Complete data sets can be found under www.math.lmu.de/~pruscha/

A.1 Hohenpeißenberg Data

Monthly temperature means in $1/10\,°C$ and **yearly** temperature means in $1/100\,°C$. The latter mean value is simply the average over the twelve monthly values (multiplied by 10). A time series plot of the yearly and of the winter means can be found in Fig. 1.2 and further analyses of these data in Fricke (2006), Pruscha (2006). In the column dcly the December value of the last year is repeated—to have the three meteorological winter months side by side. The dcly value for the year 1781 is the average of the ten Dec. values 1781–1790.

Year	dcly	jan	feb	mar	apr	may	jun	jul	aug	sep	oct	nov	dec	Tyear
1781	-18	-18	-10	24	87	122	145	154	166	126	44	15	12	723
1782	12	-10	-54	0	38	94	156	176	144	108	36	-28	-23	531
1783	-23	7	3	-4	64	108	131	163	144	118	82	12	-24	670
1784	-24	-53	-46	0	21	128	132	152	136	143	23	12	-47	501
1785	-47	6	-65	-60	13	91	117	131	131	141	60	23	-19	474
1786	-19	-1	-30	-5	71	91	139	118	123	94	35	-5	-10	517
1787	-10	-35	8	33	38	72	141	143	161	120	93	18	39	693
1788	39	-19	21	23	56	116	148	176	140	135	56	-6	-105	618

H. Pruscha, *Statistical Analysis of Climate Series*,
DOI: 10.1007/978-3-642-32084-2, © Springer-Verlag Berlin Heidelberg 2013

1789	-105	-10	-5	-34	74	131	110	147	144	110	65	4	12	623
1790	12	-6	9	16	38	120	144	135	157	109	85	28	-11	687
1791	-11	9	-18	20	89	97	130	148	164	110	72	14	10	704
1792	10	-9	-20	37	78	91	138	154	158	98	81	21	-15	677
1793	-15	-34	9	26	44	88	126	181	175	114	101	41	10	734
1794	10	-16	30	59	101	108	142	184	140	95	64	35	-25	764
1795	-25	-71	-4	16	90	109	143	122	158	138	122	-0	29	710
1796	29	61	-13	-13	44	106	132	154	145	142	58	8	-25	666
.....														.....
1990	25	17	51	55	37	120	124	151	165	104	100	18	-17	771
1991	-17	-14	-27	50	42	61	119	168	166	141	58	25	-22	639
1992	-22	-4	9	28	53	125	136	164	191	124	51	49	1	773
1993	1	22	-18	14	81	125	138	139	150	112	61	-9	18	694
1994	18	5	1	61	46	106	145	189	170	117	75	62	12	824
1995	12	-20	30	5	61	104	111	181	146	98	113	20	-18	693
1996	-18	-16	-32	-12	61	98	141	140	139	80	73	27	-21	565
1997	-21	-10	27	45	37	110	128	139	170	133	62	37	11	741
1998	11	5	34	18	64	115	149	152	161	109	73	-10	2	727
1999	2	20	-32	36	60	122	125	161	155	146	78	5	-1	729
2000	-1	-20	19	25	83	126	157	129	172	126	86	46	33	818
2001	33	-0	8	43	43	131	123	163	170	87	128	0	-37	716
2002	-37	7	32	48	56	113	167	155	154	99	76	54	12	811
2003	12	-23	-38	48	63	127	193	172	207	125	43	58	13	823
2004	13	-21	2	19	71	90	134	153	164	126	102	16	4	717
2005	4	-9	-39	23	70	113	154	157	134	132	106	23	-28	697
2006	-28	-23	-25	-3	61	109	150	198	121	155	117	64	27	793
2007	27	22	29	36	112	121	151	155	149	103	68	10	-1	796
2008	-1	25	27	16	53	127	150	155	157	102	85	36	-4	774
2009	-4	-27	-17	9	101	127	129	163	177	133	73	68	-9	773
2010	-9	-46	-12	20	71	85	143	177	146	107	68	31	-25	638

Monthly and **yearly precipitation** amounts in 1/10 mm height, the latter being the sum of the twelve monthly amounts. Once again, the Dec. value of the last year is repeated at the beginning of the next line. A time series plot of the yearly and of the winter amounts can be found in Fig. 1.4.

Year	dcly	jan	feb	mar	apr	may	jun	jul	aug	sep	oct	nov	dec	Pyear
1879	578	254	619	272	1071	1039	1009	1473	1457	1645	685	861	393	10778
1880	393	385	253	315	967	1203	1991	1870	1212	997	1784	473	907	12357
1881	907	188	232	448	809	1332	1404	885	1490	1173	828	263	202	9254
1882	202	186	88	400	696	952	1565	1802	1314	1275	788	896	501	10463
1883	501	310	127	421	504	1179	2096	2020	835	1152	526	614	684	10468
1884	684	649	202	412	1164	440	1846	1957	1130	534	1360	268	432	10394
1885	432	100	277	643	285	1185	1437	1644	925	1366	761	462	968	10053
1886	968	244	171	436	828	625	2214	972	2288	382	430	436	682	9708
1887	682	145	125	888	291	1712	435	1550	718	699	647	637	952	8799
1888	952	431	667	494	1438	690	1575	1288	1733	1887	607	211	55	11076
1889	55	174	1084	502	701	1151	1738	1408	1019	1610	678	696	260	11021
1890	260	344	137	327	718	698	1426	2045	2265	1251	708	639	107	10665
1891	107	549	142	706	807	1217	741	1947	1111	1032	373	463	619	9707
1892	619	802	628	342	1236	851	1855	1886	624	1982	1091	297	278	11872
1893	278	805	678	308	28	948	754	2495	368	827	537	898	300	8946

1894	300	159	312	570	973	1537	1000	1307	1257	906	951	137	317	9426
1895	317	351	215	552	657	1312	1386	976	1307	217	690	606	881	9150
1896	881	496	118	951	1453	2067	1275	1002	2473	1518	355	167	196	12071
...														...
1990	547	332	1062	640	1050	1617	2370	1921	1733	1183	1158	796	509	14371
1991	509	482	176	519	766	1503	1984	1216	599	861	328	694	651	9779
1992	651	88	707	1228	821	166	1557	1605	1371	713	961	1993	531	11741
1993	531	618	275	493	583	1376	1587	3465	2168	1032	767	376	719	13459
1994	719	688	328	862	1104	1073	783	1419	1201	938	497	733	836	10462
1995	836	450	526	713	1035	1070	2173	1169	2133	568	385	1059	905	12186
1996	905	138	314	588	597	1353	1018	1532	1849	1018	1082	920	417	10826
1997	417	18	585	673	934	425	1724	2404	457	319	947	189	956	9631
1998	956	398	280	1012	474	569	1389	1174	666	1632	1496	923	404	10417
1999	404	598	1187	474	927	3507	1625	1560	1194	1357	423	1341	1097	15290
2000	1097	300	802	1514	549	1537	1422	1743	2006	1413	997	551	275	13109
2001	275	681	664	1162	1107	628	2183	967	1626	1621	345	1018	773	12775
2002	773	103	685	992	723	952	1541	1509	1908	2203	821	1342	606	13385
2003	606	670	513	327	265	817	800	1504	815	450	1352	485	371	8369
2004	371	1127	431	677	547	1009	1573	1712	857	944	769	509	455	10610
2005	455	551	740	431	1191	1310	693	1840	2522	593	454	432	547	11304
2006	547	400	395	1025	1601	1524	293	2453	690	679	504	433	11120
2007	433	608	520	488	173	2377	1079	2117	2065	1627	421	726	769	12970
2008	769	414	141	694	1546	942	879	1700	1709	701	735	510	448	10419
2009	448	243	700	746	269	1421	2092	1021	737	793	796	711	751	10280
2010	751	425	425	310	389	1289	1846	1880	2424	673	624	454	696	11435

A.2 Annual Data from Five Stations

Annual temperature means (Tp) in $1/100\,°C$ and **precipitation** amounts (Pr) in $1/10\,mm$ height, at the five stations.

Aachen (A), Bremen (B), Hohenpeißenberg (H), Karlsruhe (K), Potsdam (P), for the years 1930–2008. Note that the Karlsruhe data end with the year 2008. The total average of each of the ten variables is added below.

No	Year	TpA	PrA	TpB	PrB	TpH	PrH	TpK	PrK	TpP	PrP
1	1930	1019	9525	986	6394	690	11401	1078	9884	920	6912
2	1931	898	7904	898	6384	519	10716	942	11204	807	7274
3	1932	964	7930	982	6792	634	11855	998	7116	901	4996
4	1933	918	6202	913	5355	543	13019	960	6662	802	5067
5	1934	1061	6463	1075	5540	748	10295	1107	5905	1044	4909
6	1935	976	9380	983	7240	597	11968	1063	8195	897	6172
7	1936	961	8892	900	7029	642	14187	1068	7965	887	5407
8	1937	999	7912	903	6874	688	13285	1060	7395	898	6369
9	1938	992	7461	956	7151	665	11878	1016	7934	943	5438
10	1939	972	7865	912	8102	609	15819	1008	9984	886	7208
11	1940	845	7743	721	8090	516	13464	881	8660	664	5896
12	1941	885	7188	799	7491	516	12239	893	9350	719	6618
13	1942	899	7492	789	6242	603	9145	891	7363	754	5096
14	1943	1021	7450	948	5943	737	7762	1069	6363	935	4421

Appendix A: Excerpt from Climate Data Sets

15	1944	915	8829	918	7390	566	13902	979	6460	904	5349
16	1945	1059	8164	927	7065	717	9877	1013	6719	941	6070
17	1946	967	7674	898	6068	673	10323	1002	6859	884	5394
18	1947	1056	8405	920	6615	751	8674	1032	7107	867	5873
19	1948	1056	9255	1011	6550	745	10030	1018	6840	979	6327
20	1949	1068	7201	985	7388	734	10390	1060	4854	982	5104
21	1950	983	8018	926	7786	714	9551	1019	9314	906	6023
22	1951	990	8374	938	8205	703	10955	1022	8315	939	5788
23	1952	915	9995	845	7553	616	11850	1016	8786	818	5468
24	1953	1003	6238	975	6663	713	9583	1033	5310	986	4828
25	1954	929	8598	849	7960	559	13098	973	7272	812	6961
26	1955	914	7541	842	7486	583	12927	947	7450	802	6889
27	1956	828	9378	771	8531	488	12672	871	6766	723	7206
28	1957	1000	9213	940	7799	681	11802	1033	8205	902	5700
29	1958	959	8664	891	8130	645	11714	1022	8733	856	7111
30	1959	1067	5305	983	4044	746	9100	1086	4561	958	4972
31	1960	980	9210	919	7774	651	12339	1053	7644	869	6704
32	1961	1039	10025	952	8845	767	10598	1100	7958	929	7606
33	1962	833	8303	777	7008	553	10632	936	5787	764	5135
34	1963	835	6811	773	6058	568	11614	909	5763	785	4684
35	1964	967	6991	873	5802	650	11805	1056	5065	848	5014
36	1965	894	10963	823	8748	547	15045	956	10224	800	6331
37	1966	983	11211	902	8280	668	14956	1065	8058	892	7199
38	1967	1012	7685	987	7875	689	12242	1069	7186	957	6951
39	1968	939	8012	916	7841	618	12071	988	10005	875	6172
40	1969	958	7231	875	6492	596	10675	971	7900	795	5837
41	1970	943	8414	858	7617	588	12594	986	8552	799	6255
42	1971	995	5954	957	5653	669	10475	1028	4621	921	5298
43	1972	933	6915	885	6505	638	9024	947	6602	840	4997
44	1973	975	7164	953	6277	591	11161	1008	7559	883	4919
45	1974	1006	9673	998	7846	661	13089	1087	7614	943	7155
46	1975	1019	6134	1022	6172	660	11154	1063	8203	970	4286
47	1976	1040	5405	959	5776	658	9136	1078	6483	879	3746
48	1977	1013	8171	974	6383	699	13266	1073	6999	911	6611
49	1978	939	7118	861	7258	581	13260	970	9625	847	6244
50	1979	922	8939	773	6407	633	15194	1018	6957	807	5827
51	1980	937	8423	833	6652	567	12091	968	8330	778	6469
52	1981	965	10106	869	7976	645	14493	1028	10145	868	7888
53	1982	1058	9492	934	5900	733	12165	1072	9124	962	4068
54	1983	1042	8125	966	7079	739	11375	1093	7119	978	6421
55	1984	965	9838	870	6591	620	10465	990	8294	842	5548
56	1985	882	7743	792	6953	609	11550	939	6958	809	4918
57	1986	939	9096	853	6360	643	11243	1013	9030	845	7305
58	1987	898	9541	788	6690	619	11904	978	8102	762	6913
59	1988	1049	9359	972	7352	713	13450	1114	9352	951	5679
60	1989	1118	8085	1011	6461	795	11844	1123	6323	1026	4704
61	1990	1103	7548	1026	7271	771	14371	1154	6974	1017	6789
62	1991	999	6828	909	5311	639	9779	1069	5255	895	5006
63	1992	1065	8888	1013	6917	773	11741	1142	8354	989	5689

```
64 1993  995  8770  892  9088  694 13459 1089  8469  887 6556
65 1994 1119  7675 1004  7986  824 10462 1213  8062  995 7065
66 1995 1077  7207  961  6910  693 12186 1123  9108  929 6025
67 1996  888  6266  774  4460  565 10826  970  6909  748 4460
68 1997 1059  6581  953  6209  741  9631 1103  7884  942 4951
69 1998 1051  9088  957  8931  727 10417 1131  6767  950 6243
70 1999 1105  8358 1042  5634  729 15290 1162  8416 1026 4525
71 2000 1117  9463 1038  6452  818 13109 1216  7558 1047 5863
72 2001 1048  9460  940  8396  716 12775 1129  8731  933 6700
73 2002 1113  9453  990 10617  811 13385 1170  9816  980 7890
74 2003 1103  6335  953  6143  823  8369 1184  5663  981 4189
75 2004 1033  8888  962  7104  717 10610 1113  6610  941 6330
76 2005 1072  7164  966  6777  697 11304 1119  6031  955 6342
77 2006 1116  7993 1019  5993  793 11120 1161  8506 1017 5307
78 2007 1120  9669 1054  8300  796 12970 1184  7829 1046 8259
79 2008 1048  9092 1010  6997  774 10419 1159  8332 1024 5751
-----------------------------------------------------------------
mean       993  8166  921  7012  668 11729 1043  7650  896 5920
```

A.3 Daily Data from Five Stations

Daily temperature means (Tp) in $1/10\,°C$ and **precipitation** amounts (Pr) in $1/10\,mm$ height, at the five stations

Aachen (Aa), Bremen (Br), Hohenpeißenberg (Ho), Potsdam (Po),

Würzburg (Wu),

for the years 2004–2010. The variable No denotes the calendar day in the year, running from 1 to 365. To have 365 calendar days pro each year, the 29th February 2004 and 2008 have been deleted. (For the four years 2004–2007, the daily records of Karlsruhe have also been used, but not reproduced here).

No	Day	Mo	Year	TpAa	PrAa	TpBr	PrBr	TpHo	PrHo	TpPo	PrPo	TpWu	PrWu
1	1	1	2004	-11	10	-32	0	-39	12	-31	0	-2	0
2	2	1	2004	-21	0	-13	0	-63	1	-29	0	-27	0
3	3	1	2004	-45	0	-19	0	-91	0	-48	0	-45	0
4	4	1	2004	-7	10	-14	10	-77	8	-57	2	-39	7
5	5	1	2004	32	6	-18	8	-31	31	-75	0	5	2
6	6	1	2004	58	22	20	58	-5	74	-65	100	15	43
7	7	1	2004	64	0	41	0	6	33	22	2	34	0
8	8	1	2004	48	47	26	67	24	16	14	0	14	36
9	9	1	2004	57	29	45	5	13	73	3	2	42	81
10	10	1	2004	54	3	47	7	-5	0	14	9	37	1
11	11	1	2004	92	69	66	81	43	109	29	99	65	94
12	12	1	2004	51	230	48	15	29	68	29	7	51	118
13	13	1	2004	74	60	56	103	51	153	31	60	63	110
14	14	1	2004	51	96	49	56	18	13	46	37	56	24
15	15	1	2004	34	22	42	31	-20	34	18	24	31	7

Appendix A: Excerpt from Climate Data Sets

16	16	1	2004	57	31	47	85	0	57	26	67	39	19
17	17	1	2004	42	40	42	37	-3	110	40	46	37	0
18	18	1	2004	11	0	-7	7	-20	35	-5	0	1	0
19	19	1	2004	27	274	32	106	-37	42	-7	59	-13	54
20	20	1	2004	39	10	11	0	-16	119	-1	1	16	1
21	21	1	2004	-3	0	-13	0	-66	5	-22	0	-9	0
22	22	1	2004	13	1	-17	0	-77	0	-60	0	-48	0
23	23	1	2004	32	0	7	0	-67	0	-78	0	-39	0
24	24	1	2004	28	27	2	17	-22	64	-79	1	-45	12
25	25	1	2004	31	0	27	6	-27	5	-76	5	2	4
26	26	1	2004	2	0	12	6	-38	12	-50	1	-1	6
27	27	1	2004	-10	7	-23	17	-23	31	-40	0	-17	14
28	28	1	2004	1	36	-1	15	-56	5	-32	0	-14	1
29	29	1	2004	-6	8	-10	25	-60	17	-15	12	-6	0
30	30	1	2004	9	0	18	17	-39	0	1	4	-5	0
31	31	1	2004	69	5	51	102	32	0	25	48	40	0
32	1	2	2004	100	32	83	84	56	0	78	128	97	0
33	2	2	2004	110	104	95	24	68	0	74	76	101	35
34	3	2	2004	128	0	122	2	70	0	91	44	103	0
35	4	2	2004	143	0	131	16	89	0	121	0	103	0
36	5	2	2004	131	0	129	0	105	0	130	32	115	0

. . .

330	26	11	2010	-1	0	-22	1	-41	23	-11	1	1	3
331	27	11	2010	-7	0	-52	0	-51	0	-30	0	-19	5
332	28	11	2010	-23	0	-43	0	-24	46	-31	0	-11	0
333	29	11	2010	-26	59	-9	0	-44	12	-16	0	-17	58
334	30	11	2010	-22	2	-30	0	-65	0	-44	0	-29	0
335	1	12	2010	-61	7	-73	0	-67	40	-90	70	-60	28
336	2	12	2010	-64	6	-50	12	-50	0	-100	10	-82	1
337	3	12	2010	-42	0	-41	0	-65	0	-84	0	-94	9
338	4	12	2010	-19	66	-30	27	-71	0	-70	4	-61	0
339	5	12	2010	11	109	12	2	9	38	-14	5	-17	16
340	6	12	2010	-5	0	9	0	25	127	-2	0	5	102
341	7	12	2010	-24	9	-18	0	73	5	-33	3	7	89
342	8	12	2010	-16	46	-27	24	85	69	-24	95	8	245
343	9	12	2010	5	62	-2	19	-37	20	-18	8	-8	27
344	10	12	2010	24	4	-16	81	-43	71	-25	64	4	23
345	11	12	2010	45	7	61	43	-14	16	31	48	27	13
346	12	12	2010	23	20	9	0	-22	58	3	6	24	8
347	13	12	2010	-34	21	-34	12	-81	1	-38	7	-35	0
348	14	12	2010	-32	2	-36	0	-95	26	-24	53	-53	0
349	15	12	2010	-26	15	-53	0	-100	12	-39	11	-41	12
350	16	12	2010	-13	64	-30	38	-92	2	-50	34	-54	39
351	17	12	2010	-30	2	-38	8	-54	8	-40	34	-32	1
352	18	12	2010	-40	7	-61	0	-74	0	-80	0	-64	24
353	19	12	2010	-8	160	-82	9	-8	3	-102	63	-10	56
354	20	12	2010	-39	15	-69	0	15	8	-60	12	0	90
355	21	12	2010	0	2	-105	0	38	0	-81	0	-4	65
356	22	12	2010	3	75	-45	22	68	0	-37	19	17	7
357	23	12	2010	-25	171	-17	5	100	7	-1	4	22	0
358	24	12	2010	-35	46	-25	0	-16	101	-8	100	-2	164

Appendix A: Excerpt from Climate Data Sets

359	25	12	2010	-60	4	-78	8	-75	22	-54	13	-55	28
360	26	12	2010	-12	10	-33	0	-89	1	-77	15	-93	18
361	27	12	2010	-3	0	-14	6	-65	0	-38	48	-43	0
362	28	12	2010	9	0	-43	0	-19	46	-68	0	-49	0
363	29	12	2010	16	0	-68	0	-8	15	-93	0	-60	0
364	30	12	2010	5	1	-47	9	-8	0	-97	6	-65	0
365	31	12	2010	-1	2	13	1	-30	0	3	1	-32	1

Appendix B
Some Aspects of Time Series

In the foregoing text emphasis has been placed on methods from time series analysis. In this chapter, we first present two important topics, the correlation function and the spectral density function of a time series. Statistical estimation methods for these functions are given in Sects. 3.3 and 7.1, 7.2. Then we introduce the well-known family of ARMA-models and the Box-Jenkins forecast approach.

Mathematical background material and important applications can be found in von Storch and Zwiers (1999, Chap. IV), Brockwell and Davis (2006), Falk (2011), Kreiß and Neuhaus (2006).

B.1 Auto- and Cross-Correlation Function

Let a time series $Y_t, t = 1, 2, \ldots$, be given. Assume that the expectations $\mu = \mathbb{E}(Y_t)$ and the covariances

$$\mathrm{Cov}(Y_t, Y_{t+h}) = \mathbb{E}\big((Y_t - \mu)(Y_{t+h} - \mu)\big)$$

do not depend on the time point t. Then the time series $Y_t, t = 1, 2, \ldots$, is called *stationary*. Many time series methods require stationarity. Under stationarity we can define the *auto-covariance function*

$$\gamma(h) = \mathrm{Cov}(Y_t, Y_{t+h}), \qquad \text{for all } t = 1, 2, \ldots; h = 0, 1, \ldots,$$

where $h = (t + h) - t$ is called *time lag*. In the special case $h = 0$ we have

$$\gamma(0) = \mathbb{E}(Y_t - \mu)^2 = \sigma^2, \quad \text{for all } t = 1, 2, \ldots \quad [\text{variance of } Y_t],$$

where $\sigma^2 > 0$ is always assumed. Under stationarity the auto-correlations

H. Pruscha, *Statistical Analysis of Climate Series*,
DOI: 10.1007/978-3-642-32084-2, © Springer-Verlag Berlin Heidelberg 2013

$$\rho(Y_t, Y_{t+h}) = \frac{\text{Cov}(Y_t, Y_{t+h})}{\sqrt{\text{Var}(Y_t) \cdot \text{Var}(Y_{t+h})}}, \quad t = 1, 2, \ldots, h = 0, 1, \ldots,$$

do not depend on t and are denoted by $\rho(h)$. By means of $\gamma(h)$ the *auto-correlation function* $\rho(h), h = 0, 1, \ldots$, can be written as

$$\rho(h) = \frac{\gamma(h)}{\gamma(0)}, \quad h = 0, 1, \ldots .$$

The figure shows a typical auto-correlation function over 10 time-lags. We expand the functions by $\gamma(-h) = \gamma(h)$, $\rho(-h) = \rho(h)$ in a symmetrical way. For the auto-correlation $\rho(h), h \in \mathbb{Z}$, one has

$$\rho(0) = 1, \quad -1 \le \rho(h) \le 1.$$

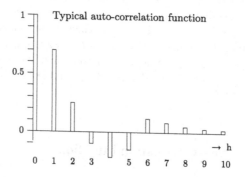

For a pure (μ, σ^2)-random series, that is a pure random series with expectation μ and variance σ^2, we have

$$\gamma(h) = \begin{cases} \sigma^2 & \text{for } h = 0 \\ 0 & \text{else} \end{cases}, \quad \rho(h) = \begin{cases} 1 & \text{for } h = 0 \\ 0 & \text{else}. \end{cases}$$

Cross-Correlation

Now, two stationary time series X_t and Y_t are given. First we have for each process

an expectation, μ_x and μ_y,
a variance, σ_x^2 and σ_y^2,
an auto-covariance function, $\gamma_{xx}(h)$ and $\gamma_{yy}(h)$, where

$$\gamma_{xx}(0) = \sigma_x^2, \quad \gamma_{yy}(0) = \sigma_y^2, \quad \gamma_{xx}(-h) = \gamma_{xx}(h), \quad \gamma_{yy}(-h) = \gamma_{yy}(h).$$

The two processes are connected by the *cross-covariance function*

$$\gamma_{xy}(h) = \text{Cov}(X_t, Y_{t+h}), \quad h \in \mathbb{Z}, \qquad \gamma_{xy}(-h) = \gamma_{yx}(h),$$

respectively by the *cross-correlation function*

$$\rho_{xy}(h) = \rho(X_t, Y_{t+h}), \quad h \in \mathbb{Z},$$

where one can write

$$\rho_{xy}(h) = \frac{\gamma_{xy}(h)}{\sigma_x \cdot \sigma_y}.$$

We have

$$\rho_{xy}(-h) = \rho_{yx}(h)$$

and $|\rho_{xy}(h)| \leq 1$.

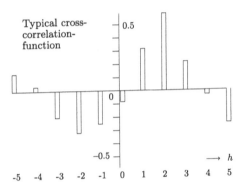

Typical cross-correlation-function

Different from the auto-correlation function, the cross-correlation function is not symmetrical and does not necessarily assume the value 1 for $h = 0$.

B.2 Spectral Density Function

By means of frequency analysis the oscillation of a time series is decomposed into harmonic components of different frequencies. The idea is that the observed series is a superposition of cyclical components with different circular frequencies ω, varying between 0 and π. Instead of ω one also uses the

frequency $\quad \nu = \omega/(2\pi)$, which lies in the interval $[0, 1/2]$,
length of period $\quad T = 1/\nu = 2\pi/\omega$, which is greater or equal to 2.

For stationary processes the most important quantity here is the *spectral density* $f(\omega)$, $0 \leq \omega \leq \pi$, also called *spectrum*. It is connected with the auto-covariance function $\gamma(h)$, $h \in \mathbb{Z}$, by the equations

$$\gamma(h) = \int_0^\pi f(\omega) \cos(h\omega) \, d\omega, \qquad h = \ldots, -1, 0, 1, \ldots \qquad \text{(B.1)}$$

$$f(\omega) = \frac{1}{\pi} \sum_{h=-\infty}^{\infty} \gamma(h) \cos(h\omega), \qquad 0 \le \omega \le \pi;$$

γ and f constitute a pair of Fourier transforms. Observe that in the special case $h = 0$ the equation

$$\gamma(0) = \sigma^2 = \int_0^\pi f(\omega) \, d\omega \qquad \text{(B.2)}$$

expresses a decomposition of the variance into the components $f(\omega)$. Due to the symmetry of the cosine function, i.e. $\cos(-x) = \cos x$, we can write $f(\omega)$ in the form

$$f(\omega) = \frac{1}{\pi} \left(\gamma(0) + 2 \sum_{h=1}^{\infty} \gamma(h) \cos(h\omega) \right).$$

We tacitly assume that $\sum_{h=1}^{\infty} |\gamma(h)| < \infty$, which is the case for the important examples of time series.

For the pure (μ, σ^2)-random series the spectral density is constant on the interval $[0, \pi]$, that is

$$f(\omega) = \frac{\sigma^2}{\pi}, \qquad 0 \le \omega \le \pi.$$

Here, all circular frequencies of the interval $[0, \pi]$ deliver the same contribution to the variance σ^2 of the time series. It is this fact, why a pure random series is called *white noise*.

B.3 ARMA Models

In this section we present an important class of time series models, the class of ARMA-models. Further, a certain variant, the ARIMA-model, is introduced. AR stands for *autoregressive*, MA for *moving average*, I for integrated. To avoid conflicts with a lower time bound, we extend the time range of a (stationary) process to $\mathbb{Z} = \{\ldots, -2, -1, 0, 1, 2, \ldots\}$.

B.3.1 Moving Average Processes

A time series Y_t, $t \in \mathbb{Z}$, is called *moving average* process of order q or MA(q)-process, if

$$Y_t = \beta_q e_{t-q} + \cdots + \beta_2 e_{t-2} + \beta_1 e_{t-1} + e_t, \qquad t \in \mathbb{Z}. \qquad \text{(B.3)}$$

Here, e_t, $t \in \mathbb{Z}$, is a pure $(0, \sigma_e^2)$-random series, $\beta_1, \beta_2, \ldots, \beta_q$ are (unknown) parameters and $q \geq 0$ is a given integer number.

An MA(q)-process is a stationary process with expectation 0. Its variance is, setting $\beta_0 = 1$,

$$\sigma^2 = \gamma(0) = \sigma_e^2 \sum_{j=0}^{q} \beta_j^2 .$$

The auto-covariance function $\gamma(h)$ and the auto-correlation function $\rho(h)$ are 0 for $|h| > q$.

MA(1): For $q = 1$ Eq. (B.3) reduces to $Y_t = \beta e_{t-1} + e_t$ for all $t \in \mathbb{Z}$. For the MA(1)-process we obtain $\sigma^2 = \gamma(0) = \sigma_e^2 (1 + \beta^2)$, as well as

$$\rho(1) = \frac{\beta}{1 + \beta^2} , \quad \rho(h) = 0 \text{ for } |h| > 1.$$

The spectral density for the MA(1)-process is

$$f(\omega) = \frac{\sigma_e^2}{\pi} \left(1 + 2\beta \cos \omega + \beta^2\right) .$$

The figure shows the (normalized) MA(1)-spectral density for $\beta = -0.5, 0, 0.5$.

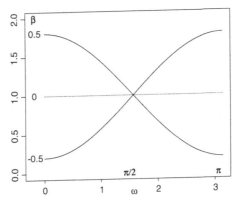

B.3.2 Autoregressive Processes

A time series Y_t, $t \in \mathbb{Z}$, is called *autoregressive* process of order p or AR(p)-process, if

$$Y_t = \alpha_p Y_{t-p} + \cdots + \alpha_2 Y_{t-2} + \alpha_1 Y_{t-1} + e_t , \qquad t \in \mathbb{Z}. \qquad \text{(B.4)}$$

Here, e_t, $t \in \mathbb{Z}$, is a pure $(0, \sigma_e^2)$-random series (further: e_t independent of Y_{t-1}, Y_{t-2}, ...). The coefficients $\alpha_1, \alpha_2, \ldots, \alpha_p$ are (unknown) parameters and $p \geq 0$ is a given integer number. An AR(p)-process is not necessarily a stationary process. Rather, this is true under the assumption, that the "stationarity condition"

the absolute values of all zeros of the polynomial

$$\alpha(s) = 1 - \alpha_1 s - \cdots - \alpha_p s^p \text{ are greater than 1}$$

is fulfilled. Under this condition the process can be represented in the form of an MA(∞)-process

$$Y_t = \sum_{j=0}^{\infty} \beta_j e_{t-j}, \tag{B.5}$$

with certain coefficients β_j [$\beta_0 = 1$]. We have $\mathbb{E}\, Y_t = 0$ for all $t \in \mathbb{Z}$, and $\sigma^2 = \sigma_e^2 \sum_{j=0}^{\infty} \beta_j^2$, with the β's from Eq. (B.5).

The first p auto-covariances and -correlations can be gained from the so-called *Yule-Walker* equations, and for $h > p$ recursively from

$$\gamma(h) = \alpha_p \gamma(h - p) + \cdots + \alpha_1 \gamma(h - 1), \tag{B.6}$$
$$\rho(h) = \alpha_p \rho(h - p) + \cdots + \alpha_1 \rho(h - 1).$$

AR(1)-Processes

For the AR(1)-process Eq. (B.4) reduces to $Y_t = \alpha Y_{t-1} + e_t, t \in \mathbb{Z}$. The stationarity condition is $-1 < \alpha < 1$. The MA(∞)-representation (B.5) of the stationary AR(1)-process amounts to

$$Y_t = \sum_{j=0}^{\infty} \alpha^j e_{t-j}.$$

The auto-covariance function and auto-correlation function are

$$\gamma(h) = \sigma_e^2 \frac{\alpha^h}{1 - \alpha^2}, \quad h = 0, 1, \ldots, \qquad \text{especially } \sigma^2 = \gamma(0) = \frac{\sigma_e^2}{1 - \alpha^2},$$
$$\rho(h) = \alpha^h, \quad h = 0, 1, \ldots \qquad (\text{therefore } \rho(h) = \alpha^{|h|} \text{ for all } h \in \mathbb{Z}).$$

As spectral density of an AR(1)-process we obtain

$$f(\omega) = \frac{1}{\pi} \cdot \frac{\sigma_e^2}{1 - 2\alpha \cos \omega + \alpha^2}. \tag{B.7}$$

For $\alpha > 0$ long-wave (low-frequency) cycles are dominant, for $\alpha < 0$ short-wave (high-frequency) cycles.

Appendix B: Some Aspects of Time Series

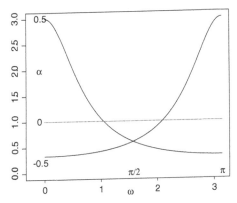

The figure shows the (normalized) AR(1)-spectral density for $\alpha = -0.5, 0, 0.5$.

AR(2)-Processes

For an AR(2)-process, that is $Y_t = \alpha_2 Y_{t-2} + \alpha_1 Y_{t-1} + e_t, t \in \mathbb{Z}$, the stationarity conditions are

$$\alpha_1 + \alpha_2 < 1, \quad -\alpha_1 + \alpha_2 < 1, \quad \alpha_2 > -1. \tag{B.8}$$

From the Yule-Walker equations we get the first two auto-correlations

$$\rho(1) = \frac{\alpha_1}{1 - \alpha_2}, \quad \rho(2) = \frac{\alpha_1^2}{1 - \alpha_2} + \alpha_2. \tag{B.9}$$

Further auto-correlation coefficients can be recursively calculated from

$$\rho(h) = \alpha_2 \rho(h-2) + \alpha_1 \rho(h-1), \quad h > 2.$$

The spectral density of the AR(2)-process is

$$f(\omega) = \frac{1}{\pi} \cdot \frac{\sigma_e^2}{1 + \alpha_1^2 - 2\alpha_1(1 - \alpha_2)\cos\omega - 2\alpha_2\cos(2\omega) + \alpha_2^2}.$$

For $\alpha_2 < 0$, more precisely $\alpha_1^2 + 4\alpha_2 < 0$, the spectral density $f(\omega)$ shows a distinct peak within the interval $(0, \pi)$.
The AR(2)-process is a suitable model for time series with a cyclic component. This maximum value of $f(\omega)$ is attained for an $\omega \in (0, \pi)$, for which

$$\cos\omega = -\frac{\alpha_1(1 - \alpha_2)}{4\alpha_2} \tag{B.10}$$

holds, under the assumption that the real number on the right side lies in the interval $(-1, 1)$.

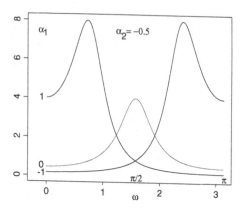

The figure shows the (normalized) AR(2)-spectral density for $\alpha_2 = -0.5$ and $\alpha_1 = -1, 0, 1$.

B.3.3 ARMA and ARIMA Processes

The combination of AR(p)- and MA(q)-terms yields an ARMA(p, q)-process. Such a process is given by the following equation,

$$Y_t = \alpha_p Y_{t-p} + \cdots + \alpha_2 Y_{t-2} + \alpha_1 Y_{t-1} + \beta_q e_{t-q} + \cdots + \beta_2 e_{t-2} + \beta_1 e_{t-1} + e_t \quad \text{(B.11)}$$

for all $t \in \mathbb{Z}$. Here we assume, that the coefficients α_i fulfill the stationarity condition. An ARMA(p, 0) [ARMA(0, q)]-process is an AR(p) [MA(q)]-process.

The first q auto-correlations $\rho(1), \ldots, \rho(q)$ depend on the α_i and on the β_j. For $h > q, p$ the $\rho(h)$ are recursively calculated acc. to (B.6), solely employing the α_i,

$$\rho(h) = \alpha_p \, \rho(h - p) + \cdots + \alpha_1 \, \rho(h - 1), \quad h > q, p.$$

ARMA(1, 1)-Processes

For an ARMA(1, 1)-process

$$Y_t = \alpha \, Y_{t-1} + \beta \, e_{t-1} + e_t, \quad t \in \mathbb{Z}, \quad |\alpha| < 1,$$

we have

$$\gamma(0) = \sigma^2 = \sigma_e^2 \, \frac{1 + 2\alpha\beta + \beta^2}{1 - \alpha^2},$$

$$\gamma(1) = \sigma_e^2 \, \frac{(1 + \alpha\beta)(\alpha + \beta)}{1 - \alpha^2}, \quad \gamma(h) = \alpha \, \gamma(h - 1), \quad \text{if } h \geq 2,$$

from where $\rho(1) = (1 + \alpha\beta)(\alpha + \beta)/(1 + 2\alpha\beta + \beta^2)$. The spectral density is

$$f(\omega) = \frac{\sigma_e^2}{\pi} \cdot \frac{1 + 2\beta\cos\omega + \beta^2}{1 - 2\alpha\cos\omega + \alpha^2} = \frac{\pi}{\sigma_e^2} \cdot f_{AR(1)}(\omega) \cdot f_{MA(1)}(\omega).$$

Differencing a Time Series

In the context of modeling and predicting, a trend in the time series $Y_1, \ldots, Y_n$ is often removed by building differences of two succeeding variables. The *differenced* time series then is $\nabla Y_2, \ldots, \nabla Y_n$, with

$$\nabla Y_t = Y_t - Y_{t-1}, \quad t = 2, \ldots, n.$$

We gain back the original time series Y_t from the differenced series ∇Y_t by means of summation ("integration"). Starting with an initial value Y_1 one recursively calculates

$$Y_2 = Y_1 + \nabla Y_2, \ldots, Y_n = Y_{n-1} + \nabla Y_n.$$

If necessary the time series ∇Y_t must be differenced once more, in order to arrive at a stationary series. Differences of order d are recursively and explicitly defined and calculated by

$$\nabla^d Y_t = \nabla(\nabla^{d-1} Y_t) = \sum_{j=0}^{d} (-1)^j \binom{d}{j} Y_{t-j}, \quad t = d+1, \ldots, n.$$

ARIMA-Processes

A time series $Y_t, t \in \mathbb{Z}$, is called an ARIMA-process of order (p, d, q) or an ARIMA(p, d, q)-process, if the process X_t of its dth differences, that is

$$X_t = \nabla^d Y_t, \quad t \in \mathbb{Z},$$

forms an ARMA(p,q)-process. An ARIMA(p, 0, q)-process is an ARMA(p, q)-process.

(i) ARIMA(1, 1, 1): $X_t = Y_t - Y_{t-1}$ forms an ARMA(1, 1)-process, i.e. X_t fulfills the equation

$$X_t = \alpha X_{t-1} + \beta e_{t-1} + e_t.$$

Then the ARIMA(1, 1, 1)-process Y_t possesses the representation

$$Y_t = \alpha_1' Y_{t-1} + \alpha_2' Y_{t-2} + \beta e_{t-1} + e_t, \quad \alpha_1' = 1 + \alpha, \alpha_2' = -\alpha.$$

Due to $\alpha_1' + \alpha_2' = 1$, the stationarity condition from B 3.2 is violated and Y_t builds no stationary ARMA(2,1)-process.

(ii) ARIMA(2,1,0)-process: $X_t = Y_t - Y_{t-1}$ is an ARMA(2,0)-process, i.e.

$$X_t = \alpha_1 X_{t-1} + \alpha_2 X_{t-2} + e_t .$$

Hence the ARIMA(2,1,0)-process Y_t has the form

$$Y_t = \alpha_1' Y_{t-1} + \alpha_2' Y_{t-2} + \alpha_3' Y_{t-3} + e_t , \quad \alpha_1' = 1 + \alpha_1, \alpha_2' = \alpha_2 - \alpha_1, \alpha_3' = -\alpha_2 .$$

As in (i) the stationarity condition is violated, because the equation $\alpha_1' + \alpha_2' + \alpha_3' = 1$ is true.

Mean Value Correction

If the stationary ARMA-process Y_t has mean value $\mu = \mathbb{E}(Y_t)$ for all $t \in \mathbb{Z}$, then we need an additional term θ_0 to write the model Eq. (B.11) in the form

$$Y_t = \alpha_p Y_{t-p} + \cdots + \alpha_2 Y_{t-2} + \alpha_1 Y_{t-1} + \theta_0 + \beta_q e_{t-q} + \cdots + \beta_2 e_{t-2} + \beta_1 e_{t-1} + e_t .$$
(B.12)

Applying $\mathbb{E}$ to both sides of (B.12), we obtain

$$\theta_0 = (1 - \alpha_1 - \cdots - \alpha_p) \cdot \mu .$$

Residuals

Let a realization $Y_1, Y_2, \ldots, Y_n$ of an ARMA(p, q)-process be given. To calculate the residuals, we rewrite Eq. (B.12) with the residual variable e_t on the left side,

$$e_t = Y_t - \left(\alpha_p Y_{t-p} + \cdots + \alpha_1 Y_{t-1} \right) - \theta_0 - \left(\beta_q e_{t-q} + \cdots + \beta_1 e_{t-1} \right) \quad \text{(B.13)}$$

for $t = 1, \ldots, n$. Here, the first q residual values e and the first p observation values Y must be predefined (e.g. by $e = 0$ and $Y = \bar{Y}$), and further residual values must be recursively gained from Eq. (B.13).
Ex. ARMA(2,2): After defining e_{-1}, e_0 and Y_{-1}, Y_0, one calculates successively

$$e_1 = Y_1 - (\alpha_2 Y_{-1} + \alpha_1 Y_0) - \theta_0 - (\beta_2 e_{-1} + \beta_1 e_0)$$
$$e_2 = Y_2 - (\alpha_2 Y_0 + \alpha_1 Y_1) - \theta_0 - (\beta_2 e_0 + \beta_1 e_1)$$
$$\ldots \ldots$$
$$e_n = Y_n - (\alpha_2 Y_{n-2} + \alpha_1 Y_{n-1}) - \theta_0 - (\beta_2 e_{n-2} + \beta_1 e_{n-1}) .$$

Residual-Sum-of-Squares. Estimation

Note that the residual variable e_t in Eq. (B.13) depends on the unknown parameters μ, α, β. In order to get estimations of these parameters, one builds the *residual-sum-of-squares*

$$S_n(\mu, \alpha, \beta) = \sum_{t=1}^{n} e_t^2, \tag{B.14}$$

and tries to find those values for the μ, α, β, which minimize (B.14) (*least-squares-* or LS-method).

B.4 Predicting in ARMA Models

General scheme. We start from an observation of a time series up to a fixed time point T, that is from the sample

$Y_1, Y_2, \ldots, Y_T.$ *[past]*

We want to make a prognosis (prediction, forecast) of future values

$Y_{T+1}, Y_{T+2}, \ldots$.   *[future]*

This prognosis is denoted by

$\hat{Y}_T(1), \hat{Y}_T(2), \ldots,$ *[forecast]*

the error of the prognosis by

$\hat{Y}_T(1) - Y_{T+1}, \hat{Y}_T(2) - Y_{T+2}, \ldots$.   *[forecast-errors]*

The function $\hat{Y}_T(l)$, $l = 1, 2, \ldots$, is called *forecast*-function at time point T for *time lead* $l = 1, 2, \ldots$.

$\ldots$	Y_{T-1}	Y_T	$\hat{Y}_T(1)$	$\hat{Y}_T(2)$	$\ldots$
$\ldots$	$T-1$	T	$T+1$	$T+2$	$\ldots$

The forecast-function is derived under the following principles:

1. $\hat{Y}_T(l)$ is a function of the observations $Y_1, Y_2, \ldots, Y_T$
2. Among all those functions, $\hat{Y}_T(l)$ is the one with the smallest mean squared error

$$\mathbb{E}\left(\hat{Y}_T(l) - Y_{T+l}\right)^2.$$

This (in the sense of 1. and 2.) best predictor for Y_{T+l} turns out to be the *conditional* expectation of Y_{T+l}, given the observations $Y_1, Y_2, \ldots, Y_T$ up to time point T,

$$\hat{Y}_T(l) = \mathbb{E}\left(Y_{T+l} | Y_1, \ldots, Y_T\right). \tag{B.15}$$

B.4.1 Box-Jenkins Forecast-Formulas

If we have a stationary ARMA(p, q)-process with mean value μ, that is cf. Eq. (B.12)

$$Y_t = \alpha_1 Y_{t-1} + \cdots + \alpha_p Y_{t-p} + \theta_0 + \beta_1 e_{t-1} + \cdots + \beta_q e_{t-q} + e_t, \quad (B.16)$$

with $\theta_0 = (1 - \alpha_1 - \cdots - \alpha_p) \cdot \mu$, then the equation for the time point $T + l$ is

$$Y_{T+l} = \alpha_1 Y_{T+l-1} + \cdots + \alpha_p Y_{T+l-p} + \theta_0 + \beta_1 e_{T+l-1} + \cdots + \beta_q e_{T+l-q} + e_{T+l}.$$
$$(B.17)$$

By building conditional expectations (B.15) on the left and right sides we get

$$\hat{Y}_T(l) = \alpha_1 \mathbb{E}_T[Y_{T+l-1}] + \cdots + \alpha_p \mathbb{E}_T[Y_{T+l-p}] + \theta_0$$
$$+ \beta_1 \mathbb{E}_T[e_{T+l-1}] + \cdots + \beta_q \mathbb{E}_T[e_{T+l-q}] + \mathbb{E}_T[e_{T+l}].$$

Hereby we have denoted, for $Z = Y$ or $Z = e$, by

$$\mathbb{E}_T[Z] = \mathbb{E}(Z|Y_1, \ldots, Y_T)$$

the conditional expectation of Z, given the observations $Y_1, \ldots, Y_T$. One determines the $\mathbb{E}_T[.]$-values according to the following scheme:

Time points up to (including) T		Time points after T	
$\mathbb{E}_T[Y_{T-j}] = Y_{T-j}$	$j \geq 0$	$\mathbb{E}_T[Y_{T+j}] = \hat{Y}_T(j)$	$j \geq 1$
$\mathbb{E}_T[e_{T-j}] = e_{T-j}$	$j \geq 0$	$\mathbb{E}_T[e_{T+j}] = 0$	$j \geq 1$

Therefore, the Box and Jenkins (1976) forecast-function can be calculated step-by-step according to Eq. (B.17), obeying the prescriptions

- at time points t up to T:
  let the variables e_t and Y_t unaltered
- at time points t after T:
  set the e_t's to zero and replace the Y_t's by their predictors $\hat{Y}$.

Hence we have for $l = 1$

$$\hat{Y}_T(1) = \alpha_1 Y_T + \cdots + \alpha_p Y_{T-p+1} + \theta_0 + \beta_1 e_T + \cdots + \beta_q e_{T-q+1}. \quad (B.18)$$

For $1 < l < p$ and $< q$:

$$\hat{Y}_T(l) = \alpha_1 \hat{Y}_T(l-1) + \cdots + \alpha_{l-1} \hat{Y}_T(1) + \alpha_l Y_T + \cdots + \alpha_p Y_{T-p+l} \quad (B.19)$$
$$+ \theta_0 + \beta_l e_T + \cdots + \beta_q e_{T-q+1}.$$

Appendix B: Some Aspects of Time Series

For $l > p$ and $> q$:

$$\hat{Y}_T(l) = \alpha_1 \, \hat{Y}_T(l-1) + \cdots + \alpha_p \, \hat{Y}_T(l-p) + \theta_0 \,, \qquad \text{(B.20)}$$

which is the AR(p)-formula, without error term and with predictors $\hat{Y}$ instead of observations Y. If an MA-term is present in Eq. (B.16), the unknown error terms $e_{T-q+1}, e_{T-q+2}, \ldots, e_T$ in (B.18) and (B.19) must be recursively calculated from $Y_1, \ldots, Y_T$ according to Eq. (B.13).

Example: ARMA(2,1)-process $Y_t = \alpha_1 \, Y_{t-1} + \alpha_2 \, Y_{t-2} + \beta \, e_{t-1} + \theta_0 + e_t$.

$$\hat{Y}_T(1) = \alpha_1 \, Y_T + \alpha_2 \, Y_{T-1} + \theta_0 + \beta \, e_T$$
$$\hat{Y}_T(2) = \alpha_1 \, \hat{Y}_T(1) + \alpha_2 \, Y_T + \theta_0$$
$$\hat{Y}_T(3) = \alpha_1 \, \hat{Y}_T(2) + \alpha_2 \, \hat{Y}_T(1) + \theta_0 \,, \text{ and so on.}$$

B.4.2 Forecast-Error and -Interval

First we need the MA(∞)-representation

$$Y_t = \sum_{j=1}^{\infty} c_j \, e_{t-j} + e_t \qquad \text{(B.21)}$$

of the stationary ARMA-process. For the AR(1)-process, e.g., we have

$$c_j = \alpha^j, \quad j \geq 1.$$

From Eq. (B.21) we gain the l-step forecast and the forecast-error,

$$\hat{Y}_T(l) = \sum_{j=l}^{\infty} c_j \, e_{T+l-j} \,, \qquad \hat{Y}_T(l) - Y_{T+l} = -\sum_{j=0}^{l-1} c_j \, e_{T+l-j} \,,$$

$[c_0 = 1]$, respectively. The expectation and the variance of the forecast-error are

$$\mathbb{E}\big(\hat{Y}_T(l) - Y_{T+l}\big) = 0 \,,$$
$$\text{Var}\big(\hat{Y}_T(l) - Y_{T+l}\big) = (1 + c_1^2 + \cdots + c_{l-1}^2) \cdot \sigma_e^2 = V(l).$$

For $l \to \infty$ the quantity $V(l)$ converges towards $\sigma_e^2 \cdot \sum_{j=0}^{\infty} c_j^2 = \sigma^2$, that is

$$V(l) \to \text{Var}(Y_t), \quad \text{if } l \to \infty.$$

A forecast-interval for Y_{T+l} at level $1 - \alpha$ has the form

$$\hat{Y}_T(l) - u_{1-\alpha/2} \cdot \sqrt{\hat{V}(l)} \leq Y_{T+l} \leq \hat{Y}_T(l) + u_{1-\alpha/2} \cdot \sqrt{\hat{V}(l)}, \qquad \text{(B.22)}$$

with $\sqrt{\hat{V}(l)} = \hat{\sigma}_e \cdot \sqrt{1 + \hat{c}_1^2 + \cdots + \hat{c}_{l-1}^2}$. Hereby, $\hat{c}_j$ and $\hat{\sigma}_e$ denote estimates for c_j and σ_e, respectively, and we have stipulated that the error variables e_t are normally distributed. $\sqrt{\hat{V}(l)}$ can be approximated by the standard deviation $\hat{\sigma}$ of the time series.

With a probability (approximately) $1 - \alpha$, a future value Y_{T+l} lies in the interval (B.22).

Appendix C
Categorical Data Analysis

The investigation of daily precipitation amounts leads us to data analysis with categorical variables. The reason is the frequent occurrence of days with amount zero (days without precipitation). If the criterion variable Y is binary, and we are interested in the dependence of Y on regressor variables $x_1, \ldots, x_m$, then the logistic regression model is often applied. If we are faced with a two-way frequency (contingency) table, then questions of independence or of homogeneity arise: Independence of the (categorical) row and column variables or homogeneity of the rows (defining certain groups).

For mathematical background material and further applications one may consult Agresti (1990), Andersen (1990), Pruscha (1996).

C.1 Binary Logistic Regression

We want to analyze a binary (dichotomous) criterion Y, assuming only the values 0 and 1, in dependence on regressors $x_1, \ldots, x_m$. Then the linear model of regression is no longer directly applicable: the range of the expectation of Y is the interval $[0, 1]$, but not the range of the linear combinations

$$\eta = \beta_0 + \beta_1 x_1 + \beta_2 x_2 + \cdots + \beta_m x_m$$

of the regressor variables. To confine the regression term to the interval $[0, 1]$, we transform η by a so-called response function $F(x), x \in \mathbb{R}$, F being monotonous and having values in the interval $[0, 1]$. We are led to the approach

$$\mathbb{E}(Y) = \mathbb{P}(Y = 1) = F(\eta).$$

Possible choices for F are the cumulative N(0,1)-distribution function (probit analysis) and the so-called logistic function (logistic regression); the latter case is outlined in the following.

H. Pruscha, *Statistical Analysis of Climate Series*,
DOI: 10.1007/978-3-642-32084-2, © Springer-Verlag Berlin Heidelberg 2013

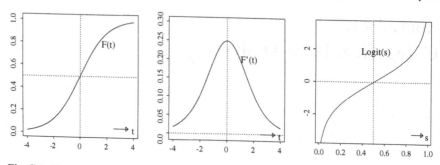

Fig. C.1 The logistic function $F(t) = e^t/(1 + e^t)$, its derivative function $F'(t) = F(t)(1 - F(t))$ and its inverse function $\text{logit}(s)$

Model and Likelihood

The *logistic* regression model uses the logistic response function

$$F(t) = \frac{e^t}{1 + e^t} = \frac{1}{1 + e^{-t}}, \quad t \in \mathbb{R},$$

see Fig. C.1, which has the so-called *logit* function as inverse,

$$\text{Logit}(s) = \ln\left(\frac{s}{1 - s}\right), \quad 0 < s < 1.$$

For the case no. i, the values of the criterion variable Y (0 or 1) and of the m regressor variables are denoted by

$$Y_i, x_{1i}, x_{2i}, \ldots, x_{mi}, \quad i = 1, \ldots, n. \tag{C.1}$$

The random variables $Y_1, \ldots, Y_n$ are presupposed to be independent. Let $\beta = (\beta_0, \beta_1, \ldots, \beta_m)$ be the p-dimensional vector of unknown parameters ($p = m + 1$), then we have for case no. i the *linear* regression term

$$\eta_i(\beta) = \beta_0 + \beta_1 x_{1i} + \cdots + \beta_m x_{mi}, \quad i = 1, \ldots, n. \tag{C.2}$$

Let us write

$$\pi_i(\beta) = P(Y_i = 1)$$

for the probability, that we will observe the event $Y_i = 1$. Then we formulate the model of binary logistic regression by

$$\pi_i(\beta) = F(\eta_i(\beta)) = \frac{1}{1 + \exp(-\eta_i(\beta))}, \quad i = 1, \ldots, n. \tag{C.3}$$

Equivalently to (C.3): The Logit(π_i), $\pi_i = \pi_i(\beta)$, is subjected to the linear equation

$$\ln\left(\frac{\pi_i}{1 - \pi_i}\right) = \eta_i(\beta) \qquad [\eta_i(\beta) \text{ as in (C.2)].} \qquad \text{(C.4)}$$

Likelihood

The unknown parameters β_j are estimated from the sample (C.1) according to the method of maximum-likelihood (ML). Starting from the

$$\text{Likelihood} \qquad \prod_{i=1}^{n}\left(\pi_i^{Y_i} \cdot (1 - \pi_i)^{(1-Y_i)}\right) \qquad \text{of the sample (C.1),}$$

we arrive—by taking logarithm—via Eq. (C.4) at the log-likelihood function

$$\ell_n = \sum_{i=1}^{n}\left(Y_i \cdot \eta_i + \ln(1 - \pi_i)\right), \qquad \text{(C.5)}$$

or, making in (C.5) the dependence on β explicit,

$$\ell_n(\beta) = \sum_{i=1}^{n}\left(Y_i \cdot \eta_i(\beta) - \ln(1 + e^{\eta_i(\beta)})\right). \qquad \text{(C.6)}$$

As estimator $\hat{\beta}$ for the parameter vector β, one chooses the ML-estimator, defined by

$$\ell_n(\hat{\beta}) = \max \ell_n(\beta),$$

where the maximum is taken over all $\beta = (\beta_0, \ldots, \beta_m)$. Plugging the estimator $\hat{\beta}$ into Eq. (C.3), we arrive at the *predicted probability* for case i, that is

$$\hat{\pi}_i = \pi_i(\hat{\beta}).$$

Classification Table

As a way to check the goodness of fit of model (C.3), we establish a so-called classification table. To this end, we choose a *cut point* c, $0 < c < 1$, and the case i is predicted (is classified as belonging) to

group 0 if $\hat{\pi}_i \leq c$ or group 1 if $\hat{\pi}_i > c$.

With respect to the actually observed value Y_i (0 or 1), this assignment can be called correct or incorrect.

In the following classification table we have $N_0 + N_1 = n$. The percentage

Observed	Classified as 0	1	Σ
$Y = 0$	N_{00}	N_{01}	N_0
$Y = 1$	N_{10}	N_{11}	N_1

$$\frac{N_{00} + N_{11}}{n} \cdot 100\,\%$$

of correctly classified cases serves us as a goodness-of-fit measure for the model. The choice of the *cut point* c: Often the value $c = 0.5$ is taken. A more appropriate choice seems to be the median $\hat{m}$ of all n values $\hat{\pi}_i$ (i.e.: 50 % of the $\hat{\pi}_i$-values are smaller (or equal) and 50 % are greater than $\hat{m}$).

More informative is a plot with two histograms of the values $\hat{\pi}_i$, separated with respect to the N_0 cases, where $Y = 0$, and the N_1 cases, where $Y = 1$; compare Figs. 6.5, 6.6.

C.2 Contingency Tables

Chi-square, Cramér's V

A contingency table consists of $I \times J$ frequencies n_{ij}, organized in a table with I rows and J columns.

	1	2	$\cdots$	J	Σ
1	n_{11}	n_{12}	$\cdots$	n_{1J}	$n_{1\bullet}$
2	n_{21}	n_{22}	$\cdots$	n_{2J}	$n_{2\bullet}$
$\vdots$	$\vdots$	$\vdots$		$\vdots$	$\vdots$
I	n_{I1}	n_{I2}	$\cdots$	n_{IJ}	$n_{I\bullet}$
Σ	$n_{\bullet 1}$	$n_{\bullet 2}$	$\cdots$	$n_{\bullet J}$	$n_{\bullet\bullet} = n$

$I \times J$-frequency table (n_{ij})

The row sums are denoted by $n_{i\bullet}$, the column sums by $n_{\bullet j}$. The total sum is $n = n_{\bullet\bullet}$.

Contingency tables arise in two different situations, which will be studied in the following under the headings "Homogeneity problem" and "Independence problem". In both cases we formulate a hypothesis H_0, namely the hypotheses of homogeneity and of independence, respectively. With the so-called expected frequencies

$$e_{ij} = \frac{n_{i\bullet} \cdot n_{\bullet j}}{n}, \tag{C.7}$$

we will employ Pearson's χ^2-test statistic

$$\hat{\chi}_n^2 = \sum_{i=1}^{I}\sum_{j=1}^{J} \frac{(n_{ij} - e_{ij})^2}{e_{ij}} = n \cdot \left(\sum_{i=1}^{I}\sum_{j=1}^{J} \frac{n_{ij}^2}{n_{i\bullet}\, n_{\bullet j}} - 1 \right) \qquad \text{(C.8)}$$

in order to check H_0. The number

$$f = (I - 1)(J - 1) \qquad \text{(C.9)}$$

is called the degrees of freedom (DF) of the table. In the following, χ_f^2 and $\chi_{f,\gamma}^2$ denote the χ^2-distribution with f degrees of freedom and its γ-quantile, respectively; see (C.11) and (C.12) below.

From the test statistic $\hat{\chi}_n^2$ we derive Cramér's V by the equation

$$V = \sqrt{\frac{\hat{\chi}_n^2}{n \cdot (K - 1)}}, \qquad K = \min(I, J). \qquad \text{(C.10)}$$

One can show that $0 \le V \le 1$ is valid.

Homogeneity Problem

Now the I alternatives, which form the rows of the contingency table (n_{ij}), represent I predefined groups. The J columns of the table stand for J alternatives, which are the possible realizations of a categorical variable Y.

In each of the I groups we have (unknown) underlying positive numbers, denoting the probabilities for the occurrence of the events $Y = j$, $j = 1, \ldots, J$. Let these probabilities in group i be

$$p_{i1}, p_{i2}, \ldots, p_{iJ} \qquad [\text{all}\, p_{ij} > 0, \ \sum_{j=1}^{J} p_{ij} = 1],$$

$i = 1, \ldots, I$. That is, we have an underlying $I \times J$-probability table; in each row of the table stands a vector with positive probabilities, adding up to 1.

	Alternative				
	1	2	...	J	Σ
Group 1	p_{11}	p_{12}	...	p_{1J}	1
Group 2	p_{21}	p_{22}	...	p_{2J}	1
⋮	⋮	⋮		⋮	⋮
Group I	p_{I1}	p_{I2}	...	p_{IJ}	1

$I \times J$-probability table (p_{ij})

The hypothesis H_0 of homogeneity asserts the equality of the I probability vectors. Under H_0, the probabilities for the single alternatives do not differ from group to group,

$$H_0: \quad p_{ij} = p_{i'j}, \quad i, i' = 1, \ldots, I, \ j = 1, \ldots, J.$$

Let the sample sizes $n_1, \ldots, n_I$ for the groups $1, \ldots, I$ be given. Assume that we have counted the frequencies

$$n_{i1}, \ldots, n_{iJ} \quad \text{in group } i \quad \left[\sum_{j=1}^{J} n_{ij} = n_{i\bullet} = n_i \right],$$

$i = 1, \ldots, I$. The set of these frequencies constitutes a contingency table (n_{ij}), with the total frequency $n = n_{\bullet\bullet}$. Using the (under H_0) expected frequencies e_{ij}, as given in (C.7), we build the test statistic $\hat{\chi}_n^2$ as in Eq. (C.8), which is under H_0 asymptotically χ_f^2-distributed. Here, f denotes the DF according to Eq. (C.9).

The hypothesis H_0 of homogeneity is rejected, if

$$\hat{\chi}_n^2 > \chi_{f,1-\alpha}^2 \tag{C.11}$$

(significance level α, n supposed to be large).

If H_0 is rejected, the question arises, which groups among the I groups are responsible. To answer this, we perform multiple comparisons between all $B = \binom{I}{2}$ pairs of two groups. The groups i, k $(i \neq k)$ differ significantly, if the test statistic $\hat{\chi}_{n_i+n_k}^2$ of the $2 \times J$ table

	1	2	... J	Σ
i	n_{i1}	n_{i2}	... n_{iJ}	$n_{i\bullet}$
k	n_{k1}	n_{k2}	... n_{kJ}	$n_{k\bullet}$
Σ	$n_{i1} + n_{k1}$	$n_{i2} + n_{k2}$	... $n_{iJ} + n_{kJ}$	$n_i + n_k$

$2 \times J$-frequency table

exceeds the quantile $\chi_{J-1,1-\beta}^2$ of the χ_{J-1}^2-distribution, where $\beta = \alpha/B$ is the Bonferroni correction of α.

Independence Problem

Now we have two variables, X and Y, where
X may assume I alternative values $i = 1, \ldots, I$
and
Y may assume J alternative values $j = 1, \ldots, J$.
Let π_{ij} the probability that we observe $X = i$ and $Y = j$,

$$\pi_{ij} = \mathbb{P}(X = i, Y = j) \quad [\text{all } \pi_{ij} > 0, \sum_{i=1}^{I} \sum_{j=1}^{J} \pi_{ij} = 1].$$

Appendix C: Categorical Data Analysis

Let a bivariate sample $(x_1, y_1), \ldots, (x_n, y_n)$ of the pair (X, Y) be given. We determine the number n_{ij} of times that the pair (i, j) occurs in this sample. This leads to an $I \times J$-contingency table (n_{ij}), as presented above.

The hypothesis H_0 asserts the independence of the variables X and Y. H_0 can be written by means of the probabilities π_{ij} and of the marginal probabilities

$$\pi_{i\bullet} = \mathbb{P}(X = i), \quad \pi_{\bullet j} = \mathbb{P}(Y = j)$$

in the form

$$H_0: \quad \pi_{ij} = \pi_{i\bullet} \cdot \pi_{\bullet j}, \quad i = 1, \ldots, I, \ j = 1, \ldots, J.$$

The Pearson test statistic $\hat{\chi}_n^2$ cf. Eq. (C.8), where the e_{ij} are once again the (under H_0) expected frequencies (C.7), is under H_0 asymptotically χ_f^2-distributed, with f being the DF acc. to Eq. (C.9). Thus H_0 is rejected, if

$$\hat{\chi}_n^2 > \chi_{f, 1-\alpha}^2 \tag{C.12}$$

(level α, n supposed to be large). From the test statistic $\hat{\chi}_n^2$ one derives Cramér's V as in Eq. (C.10). V plays the role of a correlation coefficient between the categorical variables X and Y. Indeed, the maximal value $V = 1$ is assumed, when each column (if $J \geq I$) resp. each row (if $I \geq J$) of the table (n_{ij}) contains only one single frequency greater 0 (with the rest being zero).

Remark

In the two subsections above we are faced with two different underlying situations (I univariate samples and one bivariate sample, resp.) and we have to check two different hypotheses (homogeneity and independence hypothesis, resp.). Nevertheless, we can use—in (C.11) and in (C.12)—the same procedure with the same test statistic. This fact is highly appreciated from the practitioner's point of view, since in applications the two situations often merge into each other.

References

Agresti A (1990) Categorical data analysis. Wiley, New York

Andersen EB (1990) The statistical analysis of categorical data. Springer, Berlin

Andersen PK, Borgan Ø, Gill RD, Keiding N (1993) Statistical models based on counting processes. Springer, Berlin

Attmannspacher W (1981) 200 Jahre meteorologische Beobachtungen auf dem Hohenpeißenberg 1781–1980. Bericht Nr. 155 DWD, Offenbach/m

Box GEP, Jenkins GM (1976) Time series analysis: forecasting and control, revised edition. Holden-Day, San Francisco

Brockwell PJ, Davis RA (2006) Time series: theory and methods. Springer, New York

Cox DR, Lewis PAW (1966) The statistical analysis of series of events. Methuen, London

Cryer JD, Chan KS (2008) Time series analysis. Springer, New York

Dalgaard P (2002) Introductory statistics with R. Springer, New York

Fahrmeir L, Hamerle A, Tutz G (1996) Multivariate Statistische Verfahren. DeGruyter, Berlin

Falk M (2011) A first course on time series analysis with SAS. Open Source Book, Würzburg

Fricke W (2006) Klima-Fibel Hohenpeißenberg. DWD, Offenbach/m

Grebe H (1957) Temperaturverhältnisse des Observatoriums Hohenpeißenberg. Bericht Nr. 36 DWD, Offenbach/m

Hartung J, Elpelt B (1995) Multivariate Statistik, 5th edn. Oldenbourg, München

Kreiß JP, Neuhaus G (2006) Einführung in die Zeitreihenanalyse. Springer, Berlin

Malberg H (2003) Bauernregeln, 4th edn. Springer, Berlin

Malberg H (2007) Meteorologie und Klimatologie, 5th edn. Springer, Berlin

Morrison DF (1976) Multivariate statistical analysis. McGraw-Hill, Toronto

Pruscha H (1986) A note on time series analysis of yearly temperature data. J Roy Stat Soc A 149:174–185

Pruscha H (1996) Angewandte Methoden der Mathematischen Statistik, 2nd edn. Teubner, Stuttgart

Pruscha H (2006) Statistisches Methodenbuch. Springer, Berlin

Schönwiese CD (1974) Schwankungsklimatologie im Frequenz- und Zeitbereich. Wiss. Mitt. Meteor. Inst. Univ. München 24

Schönwiese CD (1995) Klimaänderungen. Springer, Berlin

Schönwiese CD (2006) Praktische Statistik für Meteorologen und Geowissenschaftler, 4th edn. Gebr. Borntraeger, Berlin

Snyder DL (1975) Random point processes. Wiley, New York

H. Pruscha, *Statistical Analysis of Climate Series*,
DOI: 10.1007/978-3-642-32084-2, © Springer-Verlag Berlin Heidelberg 2013

Torrence C, Compo GP (1998) A practical guide to wavelet analysis. Bull Am Meteor Soc 79:61–78

von Storch H, Navarra A (eds) (1993) Analysis of climate variability. Springer, Berlin

von Storch H, Zwiers FW (1999) Statistical analysis in climate research. Cambridge University Press, Cambridge

Index

H. Pruscha, *Statistical Analysis of Climate Series*,
DOI: 10.1007/978-3-642-32084-2, © Springer-Verlag Berlin Heidelberg 2013